SUPPORTABILITY
ENGINEERING
HANDBOOK

ABOUT THE AUTHOR

James V. Jones, president of Logistics Management Associates in Irvine, California, is an internationally recognized expert on supportability engineering and integrated logistics support. He is a logistics professor at the University of Portsmouth, United Kingdom, and has been a guest lecturer at selected colleges and universities. Mr. Jones is the author of several technical reference books, including the Third Edition of *Integrated Logistics Support Handbook*, published by McGraw-Hill.

SUPPORTABILITY ENGINEERING HANDBOOK

Implementation, Measurement, and Management

James V. Jones

Logistics Management Associates
Irvine, California

SOLE LOGISTICS PRESS

New York Chicago San Francisco Lisbon London Madrid
Mexico City Milan New Delhi San Juan Seoul
Singapore Sydney Toronto

The McGraw·Hill Companies

Library of Congress Cataloging-in-Publication Data

Jones, James V.
  Supportability engineering handbook : implementation, measurement,
and management / James V. Jones.
     p.   cm.
  ISBN 13: 978-0-07-147573-0 (alk. paper)
  ISBN 10: 0-07-147573-7 (alk. paper)
   1. Syestem engineering. 2. Maintainability (Engineering) 3. Reliability
(Engineering) I. Title.
  TA168.J58   2006
  620′.0045—dc22

                                       2006046822

ISBN-13: 978-0-07-147573-0
ISBN-10: 0-07-147573-7

*The sponsoring editor for this book was Larry S. Hager, the editing supervisor
was David E. Fogarty, and the production supervisor was Pamela A. Pelton. It
was set in Times Roman by International Typesetting and Composition. The art
director for the cover was Brian Boucher.*

Printed and bound by RR Donnelley.

This book was printed on acid-free paper.

McGraw-Hill books are available at special quantity discounts to use as premiums
and sales promotions, or for use in corporate training programs. For more information, please write to the Director of Special Sales, Professional Publishing, McGraw-
Hill, Two Penn Plaza, New York, NY 10121-2298. Or contact your local bookstore.

To my family

Kim
Chris
Catherine
Jonathan
Cara Elizabeth

In loving memory of

Agnes Ferne Tate Jones
Vernon Monroe Jones

Kyung

CONTENTS

FOREWORD

Throughout the industrial, government, and academic sectors, the field of logistics is continually evolving and assuming a higher degree of importance than in the past. The complexity of systems is increasing with constantly changing requirements and the introduction of new technologies on a continuing and evolutionary basis. Frequently, the life cycles of many systems are being extended while the life cycle of individual technologies are becoming shorter. Globalization requirements and international competition are increasing significantly from year to year, and the challenges of being able to first introduce new systems into the inventory and then maintain such throughout their respective life cycles are greater than ever before. The logistics for a given system (or mix of systems) is *life-cycle* oriented, and the implementation of program-related requirements in this area necessitates a highly *interdisciplinary* approach. While the past is replete with instances where logistics requirements have been relegated primarily "downstream" in the system life cycle, often resulting in a costly approach, today's environment requires that logistics be addressed as an integral part of the system design process from the beginning.

Understanding a system's support and sustainment requirements and then successfully translating them into solid design characteristics and elements will lead the engineering team to a design that is both reflective of the operational needs of the user and affordable and sustainable. The achievement of that integration, however, is a never-ending challenge. The emphasis on sustainment of the final system design demands an understanding of the broad spectrum of logistics. That spectrum includes both consideration of reliability, maintenance and support (e.g., procurement and supply support, transportation and handling, support equipment, personnel, data/information, facilities), and familiarity with the overall system life cycle, including the design and manufacturing processes and the customer environment. The interfaces and interactions are numerous throughout the life cycle, and design and/or management decisions made in the initial design phase will have an impact on the activities and resource requirements in all of the subsequent phases of the life cycle. Thus, the systems engineer must demonstrate expertise in system design, sustainment and support, but also be conversant with the many other elements of a given program's life cycle. The engineer and logistician must be technically competent, must be knowledgeable of available design and analysis tools/models and their application, and must be able to effectively communicate with other internal project personnel, suppliers, customer contract and operational personnel across the board. Responding to this overall challenge requires that the practitioner in the field have a basic educational foundation in the principles and concepts of engineering design, sustainment, and logistics, supplemented by continuing education and training in key areas of all of the disciplines.

This First Edition of the *Supportability Engineering Handbook* constitutes a significant contribution to the understanding of the interrelationships between the design engineer and system sustainment requirements. The handbook is organized to guide the reader through the impact on design of the traditional elements of logistics which are considered during the development process. A companion to the author's *Integrated Logistics Support Handbook* (Third Edition)— which provides the fundamental introduction to the ILS challenge and fundamental definition of many of the terms used in this book—this volume builds a solid foundation for the reader's understanding the relationships of the system life-cycle phases, the life-cycle ownership considerations, and the system engineering activities during the program life cycle.

The author, James V. Jones, is an internationally recognized logistics expert and has conducted many workshops and training programs throughout the United States, Asia, Australia, and Europe. He has consulted extensively, and has an excellent reputation with defense contractors, government agencies, and throughout various facets of commercial industry. Much of his experience is reflected both in the technical content of this excellent handbook and in its organization. Jim is to be congratulated for this accomplishment.

Finally, this work is the second offering in the newly established **SOLE Press**, and is an invaluable addition to the logistics library. It serves as an excellent guide not only for the practicing logistician as well as the design and systems engineers, but also for those outside of these fields seeking additional knowledge in the impact of supportability requirements on system design. It is readily adaptable for practitioners at all levels of expertise, and the material presented allows for an easy understanding of the details of the profession.

BENJAMIN S. BLANCHARD, FELLOW, CPL
Virginia Polytechnic Institute & State University
Co-Editor, SOLE Press

ANTHONY E. TROVATO, CPL
Full Spectrum Logistics, Inc.
Co-Editor, SOLE Press

PREFACE

Supportability engineering is an internationally accepted analysis methodology and management process that today is applied to virtually every major acquisition program in both the military and civilian sectors. The guiding principles of supportability engineering form the basis for achieving the highest level of operational effectiveness and operational availability within an acceptable cost of ownership. There is a fine balance between the performance and supportability characteristics of a system that must be underpinned with a reasonable investment in both the system design and the necessary support infrastructure. Supportability engineering is in the forefront of all activities that have proven to be the best approach to achieve this balance. There is no single approach to applying supportability engineering to a systems acquisition program. The art of applying this process depends on the technology of the system, the procurement strategy, the support strategy, and the affordability of the product. There is no pattern that can be followed because every program is different. However, it has proven its worth in lowering warranty liabilities and service contract requirements and limiting the size and scope of the support footprint needed to sustain operation of the system.

This book is the result of encouragement that I have received from many colleagues, students, and attendees at workshops and seminars and the points that readers of my other books have taken the time to correspond with me about over the past few years. To these individuals, my collective appreciation and thanks.

Supportability engineering is a rewarding and satisfying profession that continually presents complex challenges. It is my hope that readers of this book will find herein some small aid in meeting and resolving these challenges.

JAMES V. JONES
Logistics Management Associates
Irvine, California

CHAPTER 1
INTRODUCTION

Supportability engineering is a technical process that focuses on the design and development of products. Before launching into a detailed introduction, two definitions are in order.

Support   The *physical act* of enabling and sustaining an item to achieve a predetermined goal or objective.

Supportability   A *prediction or measure* of the characteristics of an item that facilitate the ability to support and sustain its mission capability within a predefined environment and usage profile.

These two definitions are the foundation of supportability engineering. To understand the full scope of supportability engineering, it is easier to start with support. The definition of *support* indicates that when an item is being used, it normally requires some type of support in order to maintain its capability. For example, the use of an automobile requires many different types of support, including fuel, lubricants, and coolants while it is operating. Occasionally, the automobile breaks and requires repair. This is another type of support. By taking the automobile to a repair facility, it is returned to a completely operable state. Repair and maintenance also are support. In fact, anything that is done to service or repair the automobile can be classified as support. It is important to point out that support occurs when the item, in this example an automobile, is in the possession of the owner and is in active use. There are many issues that create requirements for support, but they can be grouped into two general categories, the design of the item and how the item is used. The type of support required for an item is precipitated by its design characteristics, and the quantity of support is related to the usage rate of the item. In this text, both the type and quantity of support will be discussed.

Supportability engineering is a relatively new engineering discipline that has developed with the evolution of technologies that created significant support challenges. Historically, products tended to be fairly noncomplex, so their support solutions typically were simple activities. However, over the past few years, the complexity of contemporary systems has created the need for a dedicated focus to ensure that the design characteristics of the system allow it to be supported. The purpose of this text is to provide a description of the processes and methods available to produce supportable systems.

GOALS OF SUPPORTABILITY ENGINEERING

The goals of supportability engineering, as shown in Figure 1-1, are focused on minimizing the cost of ownership and production through identification and elimination or control of cost drivers.

Before we discuss these goals, a discussion of the term *cost* is necessary. Various definitions of cost are provided at Figure 1-2. Typically, the word *cost* is assumed automatically to mean an expenditure of money. However, it also can have other meanings. These meanings can be even more important than an expenditure of money. The loss of a critical capability, loss of benefit, or even suffering can be far more significant than the expenditure of money. Supportability engineering must consider all these implications when addressing the design characteristics of a system. And it is the system as a whole that must be addressed, not just its component parts.

- Minimize cost of user ownership
- Minimize cost of system production
- Identification of potential cost drivers
- Participate in design and procurement decisions to control or eliminate cost drivers
- Produce a system design that can be supported using a reasonable resource package

FIGURE 1-1 Supportability engineering goals.

The goals of supportability engineering address every aspect of the design and procurement of a system. The first goal of supportability engineering looks at the long-term implications of ownership by the person or organization that will use the system. All individuals and organizations strive to receive value for systems that they purchase and use. This value is measured in various ways: satisfaction in system performance, ease of use, ease of sustaining and maintaining, and perceived value in the investment. Therefore, the goal of minimizing the cost of user ownership becomes the most important focus of supportability engineering because it contributes very significantly to the satisfaction of the user of the system.

Supportability engineering also has a significant importance to the manufacturer of the system. Through various techniques that will be discussed in this text, the system can be designed to reduce the cost of production. This leads either a lower sales price to make the product more attractive in the marketplace or produce a higher profit margin. Both of these can increase sales and therefore foster corporate profitability. An additional benefit also can be realized by manufacturers that sell products with a warranty or service agreement. The manufacturer assumes a liability to repair, service, and support the system. This liability is not only a financial liability but also a long-term liability for customer satisfaction. The concept of producing a product that will result in long-term owner or user satisfaction is a significant issue for a manufacturer to build and maintain a successful reputation for quality products.

In order to be effective in achieving these first two goals, supportability engineering must be able to identify issues that affect the cost of owning and using a system. This is accomplished through a series of interrelated analyses of the use of the system, its intended environment of use, the technology baseline of the system, and the intrinsic characteristics of system design. Each of these analysis techniques is presented in this text.

As issues that potentially increase ownership costs are identified, supportability engineering participates with other groups within the engineering activities of system design and procurement to

- The amount paid or given for something received
- Whatever must be given, sacrificed, suffered, or forgone to secure a benefit or accomplish a result
- The expenditure or outlay of money, time, or labor
- Loss, deprivation, or suffering as the necessary price of something gained or as the unavoidable result or penalty of an action

FIGURE 1-2 Definition of costs.

devise and implement solutions to eliminate, mitigate, or control their effect. This is accomplished through continuous participation in the decision-making process of system development. Some solutions are based on previous experience with similar systems, whereas others are developed through mathematical modeling and focused analysis techniques.

The success of supportability engineering is measured in the final system design solution. A system design that can be used to meet its intended purpose in a predetermined operational environment using a reasonable quantity of support resources for a reasonable and acceptable cost is considered to be a success for supportability engineering.

HISTORY OF SUPPORTABILITY ENGINEERING

Supportability engineering, as stated previously, is a relatively new discipline. Its evolution can be traced back to the 1950s through events before, during, and after World War II and the rapid development of military systems. Prior to World War II, military systems and equipments typically were mechanical and noncomplex. Then, in World War II, there was a tremendous technology growth in military systems to meet the war needs. This need precipitated development of many new and unproven technologies and their application to military requirements.

Historical records show that many "experimental" systems were developed quickly and sent into action. Some worked, and some did not. However, both experienced support problems because all development effort was focused on the performance of the systems. This caused many significant problems that had to be overcome by the military users. Shortages of spares, inadequate operation and maintenance documentation, inappropriate tools and support equipment, and incomplete training of operator and maintenance personnel were constant challenges to support the systems. Another very important point is that during World War II, the cost of developing systems was not an issue, only the performance that the systems achieved.

In the technology boom following World War II, it was realized that more importance must be placed on the support of systems as they were being developed. There had been too many failures during the war. Also, the cost of military systems in the post-World War II era became a major factor in future procurements. The military launched several initiatives in an attempt to prevent the failures and shortcomings experienced during the war. The initial activities included establishment of centers of excellence that were tasked with addressing specific areas of support, as shown in Figure 1-3. These centers of excellence established processes, policies, and procedures that greatly enhanced the capability of the military to deliver physical resources required to support systems when they were developed and placed into service.

The centers of excellence proved to be very successful in their specific area of responsibility. However, as time passed, it was realized that the centers focused on their own specialties and did

- Operation and maintenance planning
- Manpower and personnel
- Supply support
- Technical documentation
- Training
- Support and test equipment
- Facilities
- Packaging, handling, storage, and transportability

FIGURE 1-3 Support areas.

The disciplined and unified management of logistics technical disciplines to

1. Participate in the development or selection of the design of the product.
2. Plan and develop support for the product.

FIGURE 1-4   Integrated logistics support.

not coordinate the efforts into a single integrated support solution. Thus, in 1965, the military took the next logical step, which was integration of the centers of excellence. This was achieved with the creation of a unified approach to support development, *integrated logistics support* (ILS). The charter of ILS is shown in Figure 1-4. ILS is a management process that is responsible for integrating the individual logistics technical discipline centers of excellence into a single, cohesive activity to deal with the issues that historically had caused problems with support of military systems.

The goals of ILS are shown in Figure 1-5. These goals form the basis for the embryonic development of supportability engineering.

From 1965 to 1978, ILS developed into a significant participant in the acquisition of military systems. There are abundant reports of the success of ILS in developing and delivering reasonable support resource solutions for systems, the second challenge of its charter. However, ILS had only minimal success in achieving its first responsibility, participating in the design or selection of systems. ILS had its roots in developing physical logistics solutions after the design was completed. It had no experience base in dealing with design and procurement decisions pertaining to systems.

The major focus of the military is to use systems, not to acquire them. The military historically has struggled with acquisition of systems because that is not its primary function. Its organization is optimized for military operations, not acquisition. ILS also had this orientation, focusing on resources for operation rather than the design of systems. Design of systems was the responsibility of defense contractors. The military's involvement was limited to specifying a requirement that the defense contractor would design a system to meet.

Defense contractors also had implemented ILS within their organizations, but the contractors' ILS organizations also focused on identification and development of support resources after system design was completed. Therefore, the contractors' ILS organizations developed no methodology for addressing support issues. Any successes achieved by contractors in addressing support issues was through the actions of dedicated individuals rather than any widespread common methodology.

The 1970s were a time of phenomenal technological innovation. Systems moved to very complex design solutions to meet the ever-increasing needs of the military. Systems that included both

- Achieve lowest cost of ownership
- Influence design decisions for support
- Identify cost drivers
- Identify and develop support resources

FIGURE 1-5   Goals of integrated logistics support.

Functional logistics	Participation in the evolution of the design to assure it has characteristics which will allow it to be supported effectively for a reasonable ownership cost
Physical logistics	Identification, development, and delivery of the resource package required to support the system in its operational environment

FIGURE 1-6   Contemporary focuses of acquisition logistics.

hardware and software posed new support problems. The military realized that change was required if support solutions were to be effective. In meeting this need, the military looked to the commercial sector to learn how civilian industry dealt with the demands of commercial customers for supportable systems. Through a series of studies of how commercial industry solved this problem, the military developed a methodology for interjecting support considerations into the design of systems. This new methodology was called *logistics support analysis* (LSA).

In 1984, the military issued an engineering standard for incorporation of LSA into the early design and development of systems. LSA was based on best commercial practices. LSA was envisioned to be a process to be used by the ILS organization to meet its goals.

The LSA process proved to be effective, but its application was difficult because of the convoluted descriptions contained in the military standard. All too often, LSA was applied inappropriately by focusing only on the support-resource requirements for a system rather than using it as it was intended to address design issues. The LSA process became mired in costly confusion. This has been remedied today with the publication of detailed directions for its implementation. This implementation actually divides the LSA process into two distinctly different phases: activities before the design is created or procured and activities after the design solution is completed. This division has led to the development of two significant focuses during the acquisition of systems, as illustrated in Figure 1-6. Both functional and physical logistics are required to ensure a supportable system. The requirement to perform functional logistics has been the driving force in the evolution of supportability engineering.

Supportability engineering is an engineering discipline that works within the systems architecting and systems engineering activities that develop a new system. Supportability engineering is an integral part of the system engineering process that is responsible for ensuring that the final design solution contains characteristics and attributes that will allow it to be supported. The purpose of this text is to describe the activities that should be performed by supportability engineering. While the formal and organized concept of supportability engineering has been created for application on military products, it is equally applicable to any commercial product that requires support when used by consumers. As noted previously, the concept of supportability originated in the automotive, transportation, petroleum, and aviation industries. Therefore, it is equally applicable to any complex design.

PHILOSOPHY OF PERFORMANCE, SUPPORT, AND COST

In the development of a system, three key issues always must be considered. These issues are performance, support, and cost. There must be an equal balancing of performance and support for the system to be successful, unless there is an overriding requirement that pushes them out of balance. Cost should be looked on as a concern that must be addressed in the context of both performance and

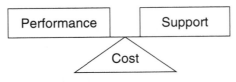

FIGURE 1-7   Balancing performance and support with cost.

support. As illustrated in Figure 1-7, there should be a balance between performance and cost that is underpinned by a reasonably acceptable cost.

Attaining a balance between performance and support underpinned by a reasonable cost is one of the basic concepts that must be understood. For example, a large, powerful engine in an automobile tends to be more expensive than a small engine. The large engine also consumes fuel at a higher rate, which makes supporting it also more expensive. The combination of cost of the engine and cost of fuel for the engine becomes a significant factor in ownership. If the consumer desires the powerful engine, then the consumer also must accept the higher fuel cost. If the consumer wants to economize on the cost of fuel, then the large engine is not as desirable as the smaller engine. Purchasing the automobile with the smaller engine will lower fuel costs, but the consumer will not enjoy the benefits of the large engine. It is the consumer's perception of value that guides the decision on which automobile to purchase. This balancing is a key issue when making the purchase decision. There is no magic solution. However, the consumer must decide prior to making the purchase. Thereafter, the consumer is committed to some level of expenditure to use the automobile, hence cost of ownership.

This concept also is used by the automotive industry when developing automobiles. Realizing that the future consumer will be astute when making purchase decisions, the industry must tailor its products to fit specific target markets. We are all familiar with the different models of automobiles offered to the public. There are typically three different ranges of products: very economical, middle range, and luxury. Very economical automobiles are for the consumer who desires a low purchase price combined with low cost for fuel and service. This consumer is prepared to forfeit comfort, power, and status to gain cheap transportation. The middle-range product is for the consumer who wants a reasonable level of comfort, power, and status but still wants to economize ownership costs. The luxury range is for the consumer who has an adequate budget to pay the penalty for maximum performance and comfort. By understanding these basic ideas, the automotive industry can produce products that fit within each target market.

The military also applies this concept to acquisition of systems. Just like the civilian consumer, the military often purchases systems that already exist. This is termed *off-the-shelf* (OTS) *procurement*. For this type of procurement, the military must make decisions in the same manner as the civilian consumer. When an OTS product is not available to meet a military need, the military must have the system designed and manufactured specifically for its requirement. Such programs are called *design programs* in this text. When the military initiates acquisition of a system using the design-program option, the military must define its requirements for performance and support. The requirements then are passed to a defense contractor who implements the military's requirements in the design solution of its product.

The design program creates the highest need for supportability engineering. As mentioned previously, acquisition and operation are distinctly different focuses of military organizations. When the military requirements are passed to a contractor, the contractor's focus is producing a product within the strict acceptance criteria established by the military. The contractor's immediate goal is to produce a system on schedule and within budget that meets the military's performance requirements. In the past, the contractor was not overly concerned with the operational support required for the system because it was delivered to the military, which would, by necessity, have to deal with that issue. This diversion of attention is what caused many of the horror stories of systems in service that were either extremely expensive or virtually impossible to support.

Supportability engineering is the discipline applied by both the military and the defense contractor as a key facet of systems engineering to achieve a balance between performance and support for a reasonable ownership cost. The military applies supportability engineering to identify system design characteristics that will allow a system to be supported for a reasonable cost. The defense contractor applies supportability engineering techniques to ensure that the appropriate design characteristics are implemented in its design solution. Both must work together to produce a cost-effective, supportable system.

OVERVIEW

The contents of this text present a logical view of the application of supportability engineering to both design and OTS procurement programs. Each subsequent chapter addresses a specific issue that must be considered in order to result in a supportable system that will minimize user cost of ownership.

Chapter 2, "System Supportability Engineering" provides an overview of system development and how systems engineering includes supportability in every aspect of product development. Of note are diagrams indicating the application of specific supportability techniques by systems engineering to achieve measurable operational outcomes through the acquisition process.

Chapter 3, "Evolving System Requirements," discusses how the true requirements for the system are identified and quantified. All requirements will be based on identification of the user's need and then defining the need in clearly understandable terms.

Chapter 4, "Creating the Design Solution," presents an overview of the system architecting and system engineering disciplines. Specific emphasis will be placed on how supportability engineering analyses are performed by systems engineering to improve availability while lowering cost of ownership.

Chapter 5, "Reliability, Maintainability, and Testability," introduces other specialty engineering disciplines that interact with supportability engineering. These disciplines produce significant information and verification of design supportability characteristics.

Chapter 6, "Supportability Characteristics," provides specific examples of the features that are included in contemporary systems to minimize user cost of ownership.

Chapter 7, "System Safety and Human Factors Engineering," includes techniques for ensuring that the system design can be operated and supported safely. It also shows how a design must consider the limitations of people who will operate and support the system.

Chapter 8, "Reliability-Centered Maintenance," describes an additional type of analysis technique that limits servicing requirements of the system.

Chapter 9, "Availability," is a detailed explanation of how all efforts during development or procurement must focus on the system being available for use and that the system must be cost-effective during use. Calculation of inherent availability (A_I), achieved availability (A_A), and operational availability (A_O) is presented.

Chapter 10, "Cost of Ownership," presents concepts for calculating the possible cost of owning a system that can be used during acquisition to attempt to limit costs of owing and using a system. Concepts of life-cycle costs (LCCs), through-life costs (TLCs), and whole-life costs (WLCs) are discussed as processes for estimating costs to assist supportability engineering in making decisions about system design and support.

Chapter 11, "Supportability Analysis," illustrates how supportability engineering performs specific analyses to identify support and cost drivers. The analysis processes then are used to develop methods of resolving the problems.

Chapter 12, "Configuration Management," addresses a critical administrative and control function that tracks the evolution of the system design solution. Configuration management documents the design baselines of the system from early functional description to the product baseline that is delivered to the user.

Chapter 13, "Supportability Assessment and Testing," provides methods for evaluating the supportability characteristics of the design solution. The methods include analyses and demonstrations.

Chapter 14, "Supportability in Service," describes how the supportability engineering process continues throughout the use of a system. This application of supportability engineering focuses on resolving design problems, participating in upgrades and modifications, and streamlining or enhancing support capabilities. A specific area of interest is dealing with obsolescence to ensure that a system can be supported adequately for its operational life.

Chapter 15, "Supportability Engineering Management," presents methods for managing the supportability engineering process to ensure that it meets its goals. Management of the process starts on day 1 of the life of a system and continues until its retirement. Proper management techniques ensure that the process meets its goals with the least amount of cost for the greatest possible benefit.

Appendix A "System Requirements Study" contains an example of this very valuable technical and management document that forms the basis for all supportability engineering activities.

Appendix B "Abbreviations and Acronyms" defines the various abbreviations and acronyms used throughout this book.

Appendix C "References" is a list of references books that can be used for further study of supportability engineering and related subjects.

Appendix D "Producibility" provides a detailed checklist for use in design and assessment of the producibility characteristics of a system prior to releasing it for manufacture.

Appendix E "Safety" contains typical contractual wording that describe how safety can be included in appropriate business agreements.

Appendix F "Human Factors Engineering" illustrates how contractual requirements can be established for human factors engineering.

Appendix G "Supportability Engineering Checklist" provides a handy list of reminders that can be used throughout the design evolution to assess the application of supportability to a system.

Appendix H "Contracting for Supportability Engineering" contains an example of the wording that may be included in appropriate business agreements to assure that supportability engineering is applied to a program.

This text has been created as a guide for successful application of supportability engineering to contemporary acquisition programs. The techniques contained herein are equally applicable to any system procurement. When applied correctly, these techniques have proven to be very beneficial in minimizing the cost of user ownership.

CHAPTER 2
SYSTEM SUPPORTABILITY ENGINEERING

Application of supportability engineering to a system is as much about timing as it is about technique. In other words, *when* something is done may be far more important than *what* is being done. For example, realizing that you bought the wrong car after you have had it for seven years is interesting but does not have value because you have had the wrong car for seven years. The same knowledge has extreme value if it can be realized before you buy the car that it is not the right one for you. Timing is more important than degree of accuracy. The realization after seven years of ownership may be factual and based on the actual expenditure of money for use and repair of the car. It would have been far better if those costs could have been estimated before you decided which car to buy and therefore avoided the wrong car. This, in a nutshell, is supportability engineering: participating in decisions about a system to achieve the lowest cost of ownership while still meeting the minimum use requirements. It is this participation in the decision-making process that is the key to success, and the decision-making process is orchestrated by systems architecting and systems engineering.

Each of these activities will be described in Chapter 3, but it is necessary to understand the concepts of system life, system ownership, and the decision-making processes before the ideas presented in this book can be applied properly. Therefore, this chapter discusses the concepts of system life cycle, system ownership life cycle, system design process, and system off-the-shelf (OTS) procurement process as a basis for application of supportability engineering. Each of the supportability engineering activities mentioned in this chapter is explained in detail in later chapters of this book.

SYSTEM LIFE CYCLE

The term *life cycle* is used to convey the idea that a system progresses through several distinct phases from its inception through use to retirement, and each phase has specific actions or conditions that bear on the utility of the system and the costs associated with its ownership. Many references are available that assign a specific name to each phase; however, a common naming convention that is universally accepted is not available. Most organizations try to assign specific actions to a phase and then link each phase together to represent the complete "life" of the system. Typically, these names might be concept, assessment, design and development, manufacturing and delivery, operation and support, and disposal, as shown in Figure 2-1. The foundation for an acquisition program is created in the concept phase. At a minimum, the user requirements document, situation and use study and organizational acquisition business plan are developed during this phase. The assessment phase consists of all activities required to determine the most appropriate approach to meeting the need defined in the user requirements document. All possible options to meet the need are identified and

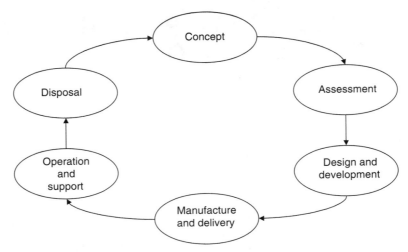

FIGURE 2-1   Traditional Life Cycle Phases.

then studied to determine the most appropriate option that gives the best potential for a reasonable balance between operational effectiveness, operational availability and cost of ownership. The option selected to meet the need is then designed, tested and baselined in the design and development phase. The manufacture and delivery phase is self-explanatory. The operation and support phase starts when the first system becomes operational and continues as long as a single system is still in service. Normally, the majority of ownership costs occur during this life-cycle phase. Disposal is the process of removing all systems from service and purging the support infrastructure of all resources that were used solely by the system. The problem with these generic phases is that one size does not fit all. Virtually every acquisition program deviates from this pattern especially off-the-shelf procurements where there is no design and development phase and the assessment phase is conducted differently. These differences are highlighted later in this chapter. So, it is best to not use these generic titles when attempting to determine when and why supportability engineering techniques are applied to a program. For the purposes of this book, names will not be given to specific phases. Phases simply will be identified by sequence numbers S1 through S7. An important issue that is overlooked by most references is how to know when a system has moved from one phase to the next. It is the identification of specific events that marks this transition. Figure 2-2 provides a depiction of the life-cycle phases for a system and indicates the phase sequence numbers, transition events, and activities within each phase.

Phase S1

The initial system life-cycle phase, S1, begins when a need or target market for a system is identified. During this phase, all activities focus on defining the requirement in measurable terms and then creating the functional solution to the system architecture. A detailed description of the application of supportability engineering process during this phase is provided in Chapter 4. The activities within this phase result in a functional description of the system and measures of success for performance, supportability, and affordability. Each chapter of this book contains pieces of the supportability engineering puzzle to be applied in phase S1. All the most significant decisions affecting cost of ownership are made during this phase. This phase ends when sufficient detail has been developed to allow a detailed physical design of the system to be started.

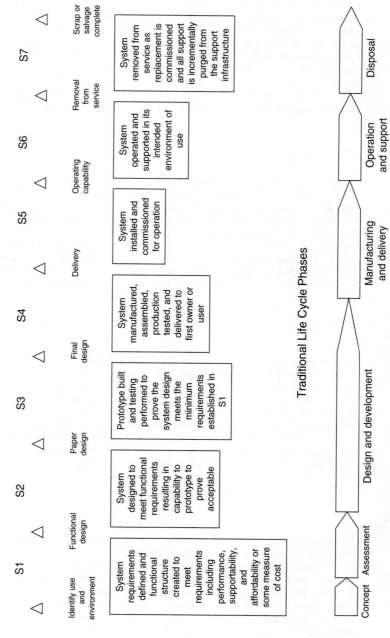

System Life Cycle Phases

S1	S2	S3	S4	S5	S6	S7

| Identify use and environment | Functional design | Paper design | Final design | Delivery | Operating capability | Removal from service | Scrap or salvage complete |

System requirements defined and functional structure created to meet requirements including performance, supportability, and affordability or some measure of cost

System designed to meet functional requirements resulting in capability to prototype to prove acceptable

Prototype built and testing performed to prove the system design meets the minimum requirements established in S1

System manufactured, assembled, production tested, and delivered to first owner or user

System installed and commissioned for operation

System operated and supported in its intended environment of use

System removed from service as replacement is commissioned and all support is incrementally purged from the support infrastructure

Traditional Life Cycle Phases

Concept — Assessment — Design and development — Manufacturing and delivery — Operation and support — Disposal

FIGURE 2-2 System life-cycle phases.

2.3

Phase S2

This phase starts with the functional system design and associated goals, thresholds, and constraints for performance, supportability, and affordability. The functional design is transformed into a "paper" design during phase S2. The term *paper design* means that it has been designed only in the abstract, and an actual physical item does not yet exist. The S2 phase starts with detailed system engineering and ends with release of the paper design produced by design engineers. Many organizations refer to the release of design as *design freeze,* indicating that no changes will be allowed without first analyzing the change and assessing its impact on performance, supportability, cost of ownership, production schedule, warranty, and any other technical, financial, or business areas. Most detailed supportability engineering analyses are performed as the paper design is created in order to avoid unwanted support liabilities when the design actually is produced.

Phase S3

The focus of phase S3 is to create from the paper design a physical item. Normally, this starts with development of a prototype that is as close to the final manufactured product as possible. The prototype is subjected to all physical and functional tests required to confirm that the design meets all the minimum requirements for performance and supportability. Some organizations have special names such as *first-article test* (FAT), *qualification test* (QualTest), *development test* (DT), *operational testing* (OT), *trials and tests* (T&T), and *field testing* to describe the extent to which the prototype will be tested to make sure that it has the highest possible probability of meeting the prescribed system requirements successfully. This phase ends with establishment of the system product, or production, baseline that will be moved into the standard manufacturing environment.

Phase S4

System manufacture and delivery occur during phase S4. Virtually no design changes are allowed during this phase because they would slow or stop the production schedule and possibly waste materials or labor, which would increase system cost. Thus all supportability engineering activities must be completed prior to the start of this phase. However, sometimes design changes occur during this phase to enhance producibility of the system. Each design change must be assessed to determine if it negatively affects supportability. Deviations and waivers discussed in Chapter 12 owing to manufacturing errors or changes in materials also must be assessed. Some degree of testing is performed during this phase, but this testing is limited to proving that each individual system manufactured continues to meet the performance levels demonstrated during phase S3.

Phase S5

Delivery of systems at the start of phase S5 normally indicates a change of ownership from the manufacturing organization to the using organization. This is not always true. For example, in the automotive industry, cars move from the producer to the seller. This is an interim change of ownership and is a significant step in the overall life of the car. Delivery of large industrial machines simply indicates that the item has arrived at its destination, where it will be installed and then used. A system may require installation, certification, training of operators, and other actions before it actually can be used for its intended purpose. At the end of this phase, the system is capable of being operated. Many organizations consider the cumulative actions taken from the start of S1 to the end of S5 to be the development of the system.

Phase S6

Phase S6 is system use. The system must be supported and maintained during this phase. All the developmental activities that occurred during phases S1 through S5 are validated by resulting in the lowest cost of ownership possible while still meeting the minimum requirements of the using organization. Normally, this phase creates the largest expenditure of funds; however, the decisions that have been made during the early phases dictate what these costs will be. This phase is also the longest phase in terms of time.

Phase S7

Phase S7 starts when removal of the system from services commences. There may be a gradual removal of individual systems from service as their replacement systems are placed into operation where multiple systems have been in use. Thus the phase may extend over a significant period of time. Also, the support resources that have been purchased and stored to support the system that have no other application must be purged from the support infrastructure. This phase ends when all systems and support resources have been discarded or disposed of as necessary to stop of costs of ownership.

This is a very simple description of the life of a system and applies to most major items from cars to aircraft to manufacturing machines. It may appear that this system life cycle is methodical and non-controversial, but every situation has variations. For example, phases S5, S6, and S7 may be repeated several times over the life of a system. The original purchaser of an item may resell the item to a second owner. The second owner later sells the same item to a third owner, and so on. A system may have several owners over its entire life. Some systems are returned to phase S4 for refurbishment before going into phase S5 again. In many cases, a system is upgraded, which means that it returns to phase S2 and then progresses through to phase S7. In all cases, a system can be tracked through its life by noting the events that mark transition between phases. There are specific supportability engineering activities in every phase.

SYSTEM OWNERSHIP LIFE CYCLE

The description of system ownership is more focused on the requirements of the system user and meeting those requirements. This life cycle, shown at Figure 2-3, is more aligned with the specific organizational structure and processes used to buy and use and item rather than the generic life of the system, as described previously. Remember, the goal of supportability engineering is to lower the ownership costs of a system, so the importance of understanding this life cycle is in pinpointing when decisions are made that result in costs.

Phase L1

The first ownership life-cycle phase begins with identification of a need or future requirement that must be met by obtaining a new system. At this point in the ownership life cycle, the focus is on defining what the need is and determining what system characteristics in terms of performance and supportability are required to meet the need. There are several options that can be pursued to meet the need. Figure 2-4 lists the possible options for obtaining a system to meet a new requirement. These include upgrading or modernizing a system that is currently in the user's possession, buying a system that already exists, or having a new system designed to meet the requirement. Each of these options has different supportability engineering issues and applications; however, all are based on producing a system that meets the need.

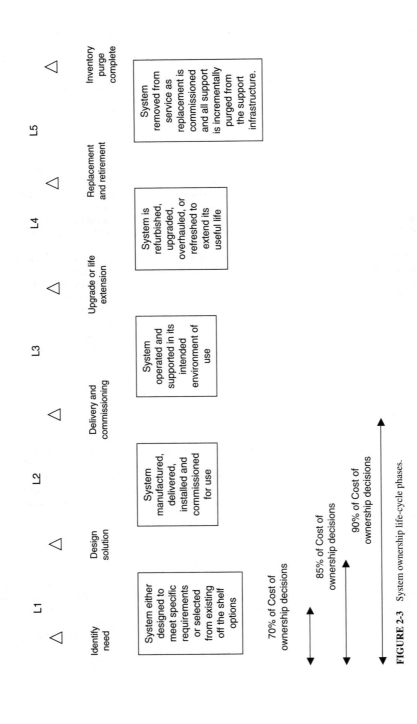

FIGURE 2-3 System ownership life-cycle phases.

1. Upgrade or modernize system currently in user's possession
2. Purchase system off the shelf
3. Modify an off the shelf system
4. Integrate existing items to create new system
5. Design new system

FIGURE 2-4 Options for obtaining systems to meet new requirements.

Phase L2

This phase starts when the design solution has been determined through one of the options listed at Figure 2-4. The system is manufactured, delivered, installed, and commissioned during this phase. Typically, each of these activities is performed in series depending on the specific requirements of the system. If a system is being purchased off the shelf, then there may be no manufacturing required. Some systems require installation before they can be used. Installation may consist of many different physical activities necessary for the system to be functional. These may include anchoring in a facility; connecting to electricity, hydraulic, or pneumatic power sources; and interfacing with other systems. There may be requirements for certification before a system can be used. Certification typically is required where legal, safety, or environmental concerns must be met. Certification also may include operation, maintenance, and supervisory personnel to ensure that the system can be operated and maintained to standard.

Phase L3

The completion of delivery and commissioning activities launches the ownership life cycle into operation of the system to meet its intended need. This is typically the longest duration of all the life-cycle phases. The majority of costs incurred owing to ownership of the system occur during this phase; however, the decisions that determine what the costs are were made in the previous phases, especially phase L1. It is extremely desirable that all benefits received from use of the system should be measured in terms of the decisions that were made during phase L1. This is necessary to assess the goodness and utility of the system. The development of measurable goals, thresholds, and constraints will be discussed in later chapters of this book.

Phase L4

Systems with extremely long ownership may require upgrading or modernization to extend their useful life. This may include upgrading the technology baseline of the system, replacing portions of the system that are obsolete, or adding new or improved functionality. Not all systems require this and jump straight to phase L5, but when they do, phase L4 signifies that the system is undergoing a significant change that will provide additional benefit for a longer period of useful life. There is an argument that the system is actually reverting back to an additional phase L1 and is going to pass through every phase again with the added functionality. This may be true conceptually; however, it is actually a smaller cycle within the larger ownership cycle. A very vivid example of this is the U.S. Air Force program for the B-52 bomber aircraft. It was conceived originally in 1945 toward the close of World War II. The B-52 first entered service 10 years later in 1955. The aircraft has been upgraded continually as missions have changed, improved technologies have evolved, and obsolescence issues have arisen. The B-52 of today is significantly different from that which was first operational in 1955. It is this continued upgrade and modernization that has kept the aircraft fully mission capable.

Phase L5

Eventually, a system must be replaced or simply outlives its usefulness. At this point, the ownership life cycle switches to activities necessary to remove the system from service. Normally, phase L5 overlaps with phases L2 and L3 of the new system replacing this aging system. This may extend over a significant time period if replacement must be done in a way so as not to degrade capability. For example, a manufacturing company may replace items of manufacturing equipment incrementally so as not to disrupt the flow of manufactured products to the marketplace. When the old system has been removed from service, then the portions of the support infrastructure that had been dedicated to the support of that system also must be purged to eliminate any continuing costs of ownership. When all systems have been removed from service and all support resources purged, then the ownership life cycle ends.

This is the point where the system life cycle illustrated in Figure 2-2 and the system ownership life cycle shown in Figure 2-3 coincide. For example, the first owner of a system sells that system to a second owner. The ownership changes from one to another. The first owner's phase L5 activities are occurring in parallel with the second owner's phase L1 and L2 activities. The first owner who had the system designed to meet a specific need, so it was a new design program. The second owner is buying an off-the-shelf (OTS) solution. This is one of the major reasons for understanding the difference between the system life cycle and the system ownership life cycle. An example that may be simpler to understand is a car. The original owner buys and uses the car for some period of time. Then the original owner sells the car to a second owner. The second owner may sell the car to a third owner and so on. The car may have many owners over its life but has only one system life cycle. For each owner there is a separate system ownership life cycle.

SUPPORTABILITY ENGINEERING APPLICATION

Virtually all supportability engineering activities should be performed during phase L1 of the system ownership life cycle, which coincides with phases S1, S2, and S3 of the system life cycle. This is when most cost-of-ownership decisions are made. This is not to suggest that there is no supportability engineering participation in later phases; however, the major benefits of the supportability engineering process can be realized only if the process is applied at the appropriate time on a program.

Design Programs

Supportability engineering activities for design programs focus on providing design engineers and purchasing specialists with explicit requirements to guide their decision making concerning the final design solution of the new system. For the purposes of this book, a *design program* is any program where the new system must be created to meet the new requirement. Options 1, 3, 4, and 5 listed in Figure 2-4 are design programs. Each of these will require preparation of a performance specification, acceptance criteria, and new items within the design. In other words, the new system does not exist and therefore must be created through the normal engineering process.

Off-the-Shelf Programs

An off-the-shelf (OTS) item is something that can be purchased and used as is without any changes. This is a normal consumer activity, where a person goes into a store to buy something off the shelf and then takes it home and uses it. The person has no direct participation in the design or development of the item. It is the person's responsibility to be an astute consumer and select the most appropriate item in the marketplace. Only option 2 listed at Figure 2-4 requires no design activities.

Commercial off the shelf (COTS)

- Built to no specific user need
- Generic environment of use
- External configuration control
- Limited detailed statistics available
- Limited flexibility in support options
- Limited access to design information
- Tested to limit warranty and liability obligations
- Mass produced so price should be competitive

Nondevelopment item (NDI)

- Built to meet a specific requirement
- Specific environment of use
- Internal configuration control
- Extensive detailed statistics possibly available
- Possible to modify support methods
- Normally full access to design information
- Tested to meet specific performance and support parameters
- Limited production so price may be high

FIGURE 2-5   Off-the-shelf items.

Therefore, no decisions are being made about the characteristics of the design, so there can be no supportability engineering activity on this type of program to influence design characteristics. Those decisions were made for the original buyer of the system or the producer that is trying to market the product. OTS items can be divided into two categories with distinctly different criteria: items that were developed for the commercial market to be purchased by any consumer and items that were purpose-built for an individual or specific organization that now can be purchased by someone else. Commercially available OTS items typically are called *commercial off the shelf* (COTS). Items that are available that were built originally for a previous owner may be called *nondevelopment items* (NDIs), *government off the shelf* (GOTS), *military off the shelf* (MOTS), or some other name to denote the specific original user of the item. Figure 2-5 shows these items and the issues that must be considered when choosing the type of OTS product to consider. The differences between COTS and NDIs are extremely significant and bear heavily on the decision-making process as to which type of item to purchase off the shelf.

SUPPORTABILITY ENGINEERING: DESIGN PROGRAMS

Application of supportability engineering techniques on a design program can be linked to evolution of the design of the item from identification of need until the design has been accepted as meeting the specified requirements of the user. There are six distinct steps on design programs, and they

---

1. Buyer issues request for proposal
2. Potential sellers submit proposals
3. Buyer selects preferred seller proposal and awards contract
4. Preliminary design review
5. Critical design review
6. Functional configuration audit and physical configuration audit

---

FIGURE 2-6 Design evolution steps.

are illustrated in Figure 2-6. There are specific supportability engineering activities that must be applied between each of these steps to ensure that the necessary supportability characteristics are in the final system design solution. Figure 2-7 provides an overview of these activities. The time between each step in the design process provides windows of opportunity to interject supportability concerns into the evolving design solution. For the purposes of this book, these windows will be referred to in the context of performing specific activities for specific purposes during each step of development.

Window D1

Window of opportunity D1 starts when the need has been identified and ends with issue of the document that invites possible sellers to submit their proposal for a product design to meet the need. This ending document typically is called a *request for proposal* (RFP), *request for quotation* (RFQ), *invitation to tender* (ITT), *request for tender* (RFT), or some similar title. The focus of D1 is to ensure that all necessary supportability requirements are included in this document. Figure 2-8 shows typical supportability engineering activities during window D1.

Window D2

This window opens when the RFP is issued and closes when sellers submit their proposals. It is of extreme importance that the proposals submitted by potential sellers include all possible solutions to supportability issues contained in the RFP. Figure 2-9 shows the supportability engineering activities that should be performed by the seller in preparation of the proposal.

Window D3

The buyer must select the proposed system design that meets the best balance between performance support and costs of ownership. The activities in window D3 provide the capability to make this decision. The first issue is to make sure that a proposed design solution has the minimum technical performance required to meet the user's needs. Then it must be determined which proposed design achieves the most reasonable balance for supportability and cost of ownership. Figure 2-10 shows the typical supportability engineering activities performed by the buyer during D3 to achieve these objectives. All measurable supportability requirements must be included in the final contract between the buyer and the seller.

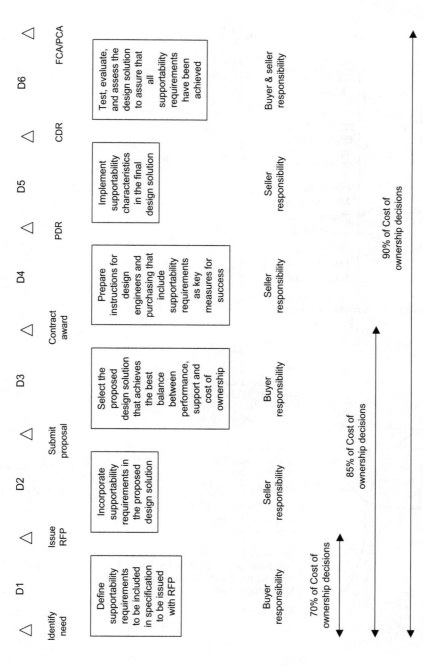

FIGURE 2-7 Supportability engineering: design program.

2.11

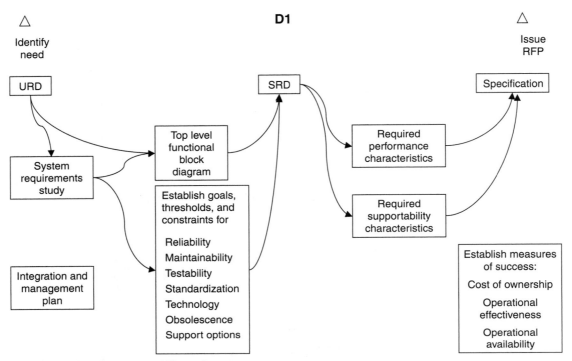

FIGURE 2-8 Establishing supportability requirements.

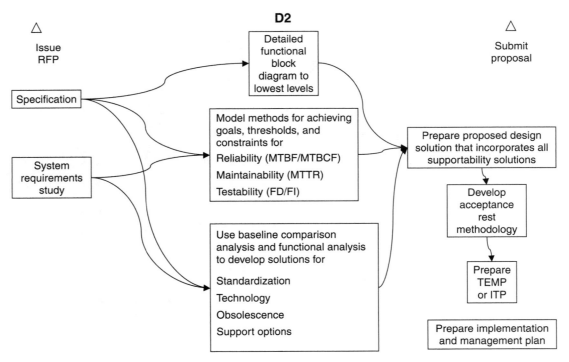

FIGURE 2-9 Developing supportability solutions.

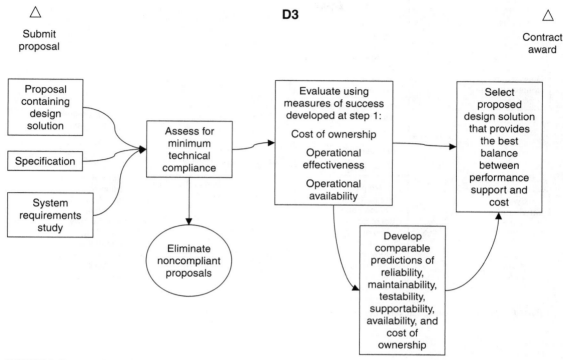

FIGURE 2-10 Assessing supportability potential.

Window D4

The purpose of supportability engineering during window D4 is to participate in the systems engineering process that takes the system-level requirements and translates them into instructions for every level of item within the total design solution. Systems engineering prepares instructions for design engineers, subcontractors, and purchasing specialists. These instructions are the basis for creating the final design solution. All supportability requirements must be included in these instructions. The activities in D4 are those that tend to be the least performed on programs until it is too late to effectively interject supportability into the design solution without significant redesign efforts. Figure 2-11 shows the actions of supportability engineering as part of the overall systems engineering process to achieve a supportable design. The results of this activity are presented at the *preliminary design review* (PDR).

Window D5

Implementation of supportability into the design solution rests with the design engineers and purchasing specialists who create the final design. They follow the instructions developed during D4. The results of their efforts are presented at *critical design review* (CDR). Supportability must be viewed as being equal in importance to performance at CDR. Figure 2-12 provides a description of all the supportability engineering activities expected to be performed during window D5. At the end of window D5, the "paper" design should be completed. This means that there may not yet be any actual physical system, only the documentation necessary to manufacture it. Thus many of the supportability engineering results are documented in predictions or confirmed through analysis of the paper design rather than the physical design.

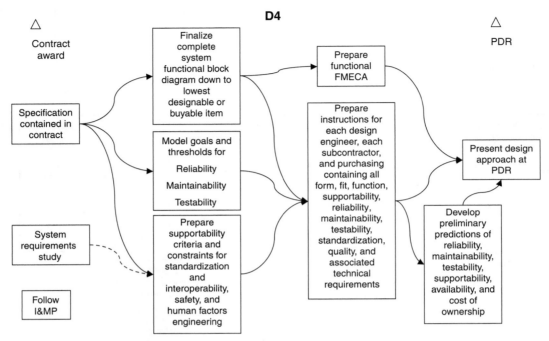

FIGURE 2-11 Implementing supportability requirements.

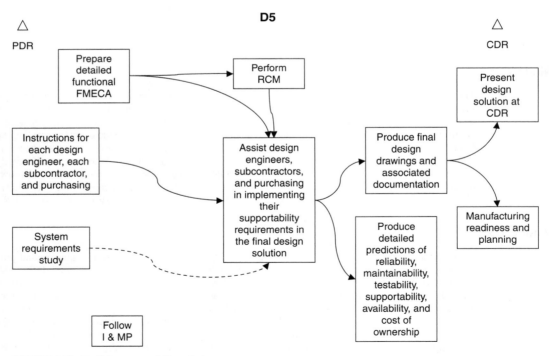

FIGURE 2-12 Realizing supportability solutions.

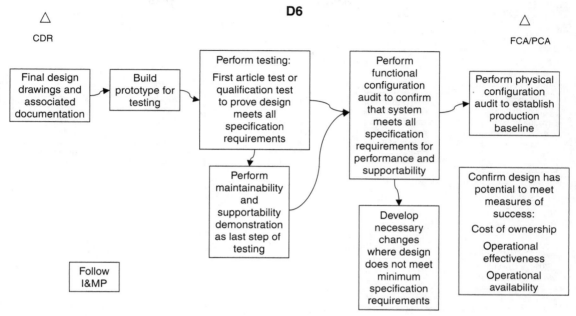

FIGURE 2-13 Demonstrating supportability requirements achieved.

Window D6

All activities performed during window D6 are to prove that the new system design meets the minimum requirements contained in the specification. This is normally done by the seller building a prototype and then subjecting the prototype to whatever testing is necessary to prove that it meets all specification requirements. The prototype may be production system serial number 1, or it may be preproduction that is built using model-shop techniques rather than standard manufacturing processes. Figure 2-13 shows that window D6 is completed when the results of all testing are assessed by functional configuration audit to confirm that the system meets specification requirements and then physical configuration audit to confirm that all documentation for the system is accurate.

Each of the supportability engineering activities performed during system development must be performed in the appropriate window of opportunity to receive its complete value. Activities that are performed in the wrong window of opportunity may be a total waste of time, or they may create more confusion than present solutions. A system should be very supportable and cost-effective to own when these activities are performed properly and in the right window of opportunity.

SUPPORTABILITY ENGINEERING: OFF-THE-SHELF PROGRAMS

Application of supportability engineering on OTS programs is different than for design programs. Supportability engineering as applied to OTS programs is a critical participant in the decision as to which OTS product to buy. All applications must be done before the final decision as to which product to buy has been made. After that decision is made, virtually all benefits that could be received from supportability engineering are past. Timing of application is even more important on OTS programs because there is no opportunity to back up and repeat decisions without canceling previous

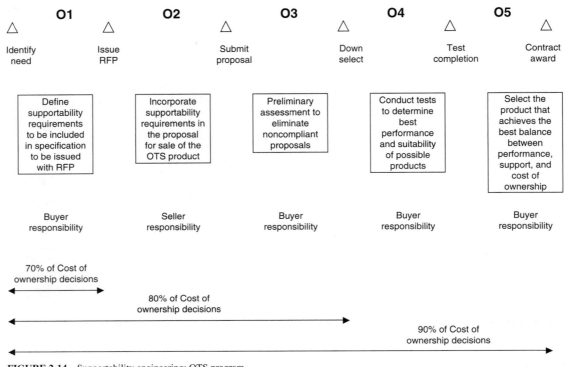

FIGURE 2-14   Supportability engineering: OTS program.

activities. Figure 2-14 provides an overview of how supportability engineering should be applied in the five windows of opportunity, windows O1 through O5, of an OTS procurement program.

Window O1

The activities performed by supportability engineering on an OTS program are very similar to those performed in window D1 of a design program. The slight variation in this window is to develop criteria that eventually will be used during the selection of which OTS system to purchase. Figure 2-15 shows the detailed activities during window O1 and how they culminate in criteria that will be used to purchase the new system. These criteria are different from the specification resulting from D1. They tend to be at a higher level and treat the system as a whole rather than trying to determine the parts of the system on a design program.

Window O2

This window starts with the buyer issuing the performance and supportability criteria that will be used to select the OTS system to be purchased. All interested sellers prepare their proposals during window O2. Each proposal must describe how the seller's OTS system that is offered for purchase will meet each of the measurable criteria. Figure 2-16 shows the actions that a seller should

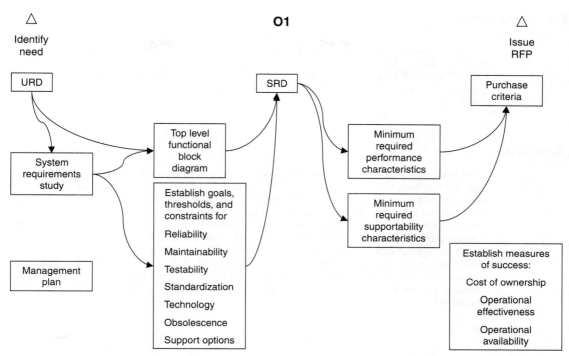

FIGURE 2-15 Establishing supportability requirements.

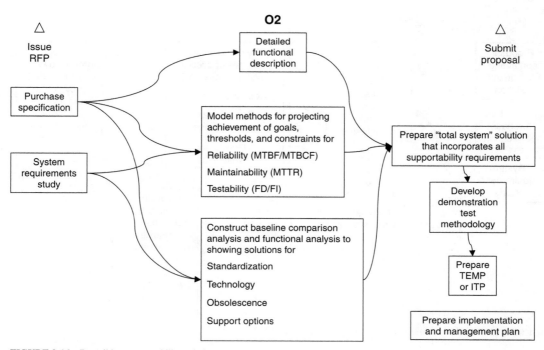

FIGURE 2-16 Describing supportability solutions.

take to prepare a proposal. It is assumed that the OTS system has been used by others, so real-world use information should be available for the seller to provide realistic projections of how the OTS system will meet the user's needs. This is one of the big advantages of an OTS program because on a design program, there is no real-world usage information that can be examined during selection.

Window O3

The buyer received a proposal from all potential sellers and performs an analysis of each OTS system that has been offered for purchase. The buyer first assesses the proposed OTS systems for minimum technical compliance with the purchase criteria developed in O1. Any proposed system that does not meet the minimum criteria is eliminated from further consideration. The OTS systems that do meet technical criteria are further assessed to determine which ones appear to provide a reasonable balance between performance supportability and cost of ownership. Figure 2-17 shows the typical methods used for this assessment. This assessment uses the information submitted by sellers with their proposals. Window O3 results in identification of the top candidates for purchase. This is sometimes called *short listing,* meaning that the list of possible systems to purchase has been reduced to a final few that will be tested to see which one ultimately will be purchased.

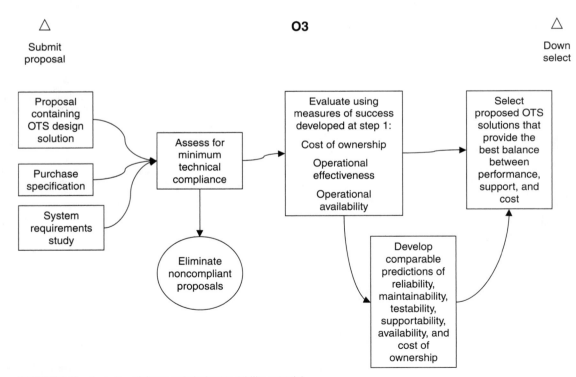

FIGURE 2-17   Assessing minimum technical supportability potential.

Window O4

Physical and functional testing is performed during window O4 to determine the actual characteristics of the OTS systems being considered for purchase. One of the key issues in selecting which OTS system to purchase is the system's suitability in the user's operational environment and mission scenario. A very good product for one user may be totally useless for another. The purpose of the testing performed during O4 is to prove or disprove the suitability of each proposed OTS system using the criteria that were developed during window O1. This testing must involve the user and should include operational testing, field testing, performance testing, supportability testing, ease-of-maintenance assessment, and support-infrastructure assessment. Every test should be performed in accordance with the test and evaluation management plan (TEMP) and the integrated test plan (ITP). Both of these are described in Chapter 13. The results of every test must be documented accurately. Figure 2-18 provides a description of this process.

Window O5

The final purchase decision is made at the end of window O5. The results of all testing performed in window O4 are compared with the purchase criteria from window O1. Extensive modeling may be required to quantify which OTS system provides the most reasonable balance among performance, supportability, and cost of ownership. Figure 2-19 shows how supportability engineering participates in this assessment process. Operational effectiveness, operational suitability, operational availability, and cost of ownership provide the final evaluation measures in this decision-making process.

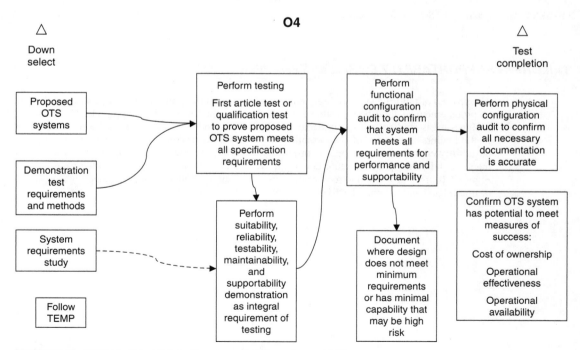

FIGURE 2-18   Testing to demonstrate performance, suitability, and supportability.

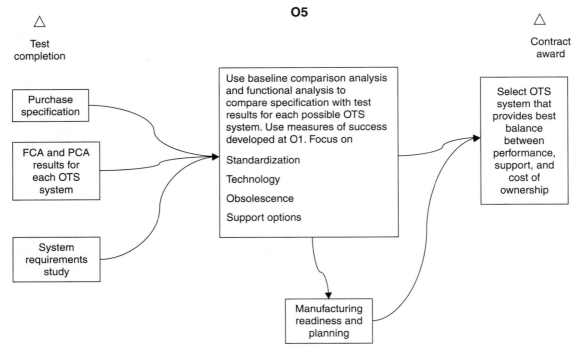

FIGURE 2-19   Selecting the OTS system for procurement.

TAILORING SUPPORTABILITY ENGINEERING APPLICATION

The process of tailoring supportability engineering to a specific program depends totally on the type of procurement method being followed. It is either applied to interject supportability into a new design, or it is applied to assist in deciding which OTS product to buy. The choice of procurement method tailors application of supportability engineering. The second and final step of tailoring supportability engineering to a specific program is simply determining the scope of effort needed to meet the individual program requirements; however, all activities contained in either Figure 2-7 or Figure 2-14 must be applied to receive the maximum benefit. Supportability engineering is a process of interrelated and integrated activities. Therefore, all activities apply.

CHAPTER 3
EVOLVING SYSTEM REQUIREMENTS

Every acquisition begins with identification of clear and understandable requirements for the system. This identification includes the intended use of the system, the environment where the system will be used, the rate of usage, concerns pertaining to support of the system, and potential costs of ownership. The importance of this identification cannot be overemphasized. Failure to perform this identification properly almost always will result in an inappropriate system solution. Therefore, due diligence must be applied to this initial process. It is the first step toward success of any acquisition program.

DEFINING THE NEED

The first step in identification of system requirements is to define the need for the system. Not what *is* the system, but *what does it have to do?* Figure 3-1 illustrates the questions that must be answered as the initial step in defining need. The answers to these questions form the basis for applying supportability engineering analyses and techniques to the acquisition and management of the system throughout its life. This list of questions represents the minimum input information necessary to establish the basis parameters that delineate the ultimate criteria for system success. Throughout this text, this list will be referenced continually because the answers to these questions represent the final acceptability of the system to meet the user's need.

Answering the questions posed at Figure 3-1 is not a simple activity. Typically, no single person or organization has all the correct answers. Experience has shown that most people or organizations have guesses or assumptions that they use as answers, but different organizations working on the project have made different guesses or assumptions that they do not share or compare. This results in confusion of project goals and parameters. The answers developed on a project normally result from coordination among all the stakeholders, arbitration where differences occur, and finally, agreement of all parties.

The answers to each question are important; however, it is the first question—addressing how the system will be used—that is most important. The actual use of the system creates requirements for support and therefore is the focus of supportability engineering. Let's look at two examples.

The cost of personnel is the major expense for the U.S. Postal Service (USPS), so the service has developed automated sorting machines to reduce personnel requirements and increase the speed of sorting mail. The concept of this operation is that all mail collected at mail boxes and postal drop-off points is consolidated and shipped to the sorting plant. The first activity is to divide mail into specific categories such as letters, small packages, priority mail, large packages, etc. Each category of mail is handled differently because of the physical requirements for handling and packaging. Letters arrive at the sorting plant in a massive jumble and leave the sorting plant in boxes ready for letter carriers to take on their delivery routes. Inside the sorting plant, there are fours steps to sorting. First, the jumble must be organized so that every letter is facing the same direction with the stamp in the upper right corner. Then the stamp is canceled. Next, ZIP codes are read. And finally, the letters are sorted into the proper

- How will the users actually use the system?
- Is there a measurable output to be produced from system use?
- What are the minimum performance requirements?
- Where will the system be used?
- How frequently will the system be used?
- Under what circumstances will the system be used?
- How will the user measure system success?
- How will the user measure system failure?
- Are there any limitations on system use?
- Are there any limitations on system characteristics?
- Are there any economic issues or constraints?
- Are there any environmental issues or constraints?
- Are there any regulatory issues or constraints?
- Are there any support limitations?

FIGURE 3-1   Critical starting questions.

destination sequence. Figure 3-2 illustrates this process. Each of the second-tier blocks in this figure represent specific machines that make up the overall sorting capability. In later chapters this illustration will be referred to as a *functional block diagram*. Each tier of blocks shown at Figure 3-2 consists of individual machines, so each machine must be developed to perform the specific function. All the individual machines must work together to meet the overall requirements of the sorting plan. The capability of the sorting plant is limited to the slowest throughout of the individual machines. The functional block diagram allows overall capability of any system to be understood visually. The functional block diagram also allows easy understanding of changes to system functionality when required.

In 2001, the USPS was the target of deadly attacks by someone putting anthrax spores in letters. This became a critical issue for enhancement of all sorting plants to identify and eliminate any subsequent attacks. Thus an additional function was added to each sorting plant. Figure 3-3 shows that a function was added to detect any possible biological hazards. It is important to note that the new function

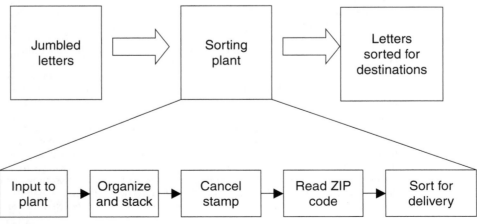

FIGURE 3-2   USPS sorting plant operations.

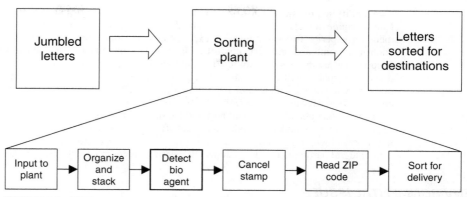

FIGURE 3-3 USPS sorting plant operations enhanced with a detector.

was added to the work flow at the earliest possible step of the overall process to limit possible contamination of the sorting plant.

The overall need for the sorting plant and each individual machine within the plant can be stated as: "The sorting plant must be capable of sorting 730,000 letters per day or 1.9 billion letters per year." This is the measurable output statistic for the system and establishes a minimum performance requirement. If the sorting plant cannot process these quantities, then mail backlog builds, and consumer satisfaction erodes. Everything that is done during acquisition of the machines within the sorting plant must enable it to achieve this requirement. There are many other facts that also must be addressed; one of the most important is the cost of the individual machines and the cost of operating the overall plant. These costs are amortized into the cost of each letter sorted by the machines. The sorting plant must operate efficiently. Any defect that stops its operation creates a break in the flow of mail. Therefore, each machine must be very reliable and also easy to restore to an operable condition when a defect occurs. Supportability engineering focuses its activities on the cost of the sorting plant, the machines within the plant, and the ability to support them.

The second example is a main battle tank used by the Army. A tank is also a very complex and expensive system. Its purpose is to be capable of defeating opposing forces, especially an opposing force's tanks. The requirements for the main battle tank therefore focus on the capabilities of any opposing forces and attributes that the tank must have to overcome any opposition. This situation is quite different from the transfer press of the automotive industry. The desired output is not as explicitly measurable in terms of production quantities. Definition of the need for the tank must start with a study of the tactical mission that the tank must perform to defeat opposing forces. The tank potentially will meet several different types of opposing-force equipment on the battlefield. Such equipment may include tanks, infantry, aviation, artillery, and antitank mines. These are commonly grouped into potential "threats" to its survival and mission success. The need for the tank can be summarized as "the tank will be capable of closing with and defeating opposing threat forces through mobility and application superior of firepower." This brief description of the need for the tank conveys the requirement for the tank to be highly mobile with an accurate and lethal gunnery system. Figure 3-4 shows these high-level requirements for the tank.

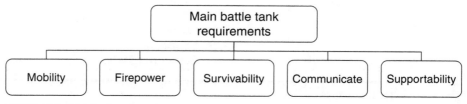

FIGURE 3-4 Main battle tank requirements.

Mobility requirements must be defined further to determine distance, time, and terrain and will lead to engine horsepower, transmission, track suspension system, and driving controls. The requirement for superior firepower must be expanded to consider munitions, cannon, sight system, fire-control system, and gunnery controls. Underlying the obvious requirements for mobility and firepower is another—survivability. The tank must be capable of surviving a combat situation. This mandatory characteristic will define the size, shape, and construction of the tank body. As the tank body increases in size, its weight also increases. An increase in weight increases the requirement for engine horsepower to improve mobility, etc. Each of these requirements is interrelated.

There is one other requirement required for operation, and that is to communicate. An unspoken requirement for the tank is its ability to operate for some period of time without failing. A tank should never breakdown in the middle of a combat mission. This leads to supportability engineering, which is an integral participant in the systems engineering process that will develop the tank design.

ENVIRONMENTAL ISSUES

The environment where the system is to be used must be described explicitly. The specific conditions to which the system will be subjected have a very significant bearing on its operability. Different environments pose different stresses on a system. Environmental issues can be divided into two categories: physical environmental effects on the system and system effects on the environment. Both theses issues must be considered prior to start of development or procurement of a system.

Physical Environment

The physical environment where a system will be used can have a significant effect on its ability to operate efficiently. Figure 3-5 lists the typical areas that must be considered. The temperature range of the location where the system will be operated has a direct impact on its success. If the location is very hot, the system probably will require some type of cooling, or if it is very cold, the system may require heating. If there are many locations where the system will be used, then it may require both heating and cooling with a capability to monitor and adjust the temperature. Humidity of the environment of operation also must be considered. High levels of humidity lead to corrosion of metal components and interfere with electronic assemblies. Extremely low humidity causes accelerated deterioration of materials such as rubber gaskets and adhesives. Airborne particles can pose additional concerns for proper system operation. Particles such as dust may require a filtration element to remove them. Systems that will operate on or near the sea are subjected to salt fog, which also can damage components. This requires a protective coating to limit its effects. Aviation systems that may operate at extreme altitudes require specific additional characteristics to accommodate the environment. Some systems may be sensitive to vibration. Violent shaking or long-term minor movement may degrade or halt system operation. Electronic systems can be affected by electromagnetic interference produced by emissions from other nearby electronic systems. The system being produced or acquired must be capable of operating within the limitations of its intended environment of use. Therefore, every possible influence present

- Temperature
- Humidity
- Air
- Altitude
- Vibration
- Electromagnetics

FIGURE 3-5 Physical environmental issues.

- Solid waste
- Liquid waste
- Gaseous waste
- Particle emissions
- Electromagnetics

FIGURE 3-6   System effect on the environment.

in this environment must be identified and analyzed to ensure that the system will be successful by eliminating the effect through design characteristics or control measures.

System Effect on the Environment

The effect of the system also must be considered at the start of procurement or development. Many systems produce by-products that can harm the environment. Examples of these are listed at Figure 3-6. A system may produce solid, liquid, or gaseous waste. These types of waste require control, storage, and disposal. Storage and disposal of environmentally hazardous waste increase the cost of system operation and support, which could cause a significant increase in ownership cost. Some systems produce particle emissions such as fumes, smoke, dust, or noxious odors that may have a negative effect on surrounding areas. There may be some by-products that are neither hazardous nor harmful but constitute a blight on the surrounding area. Electronic systems may produce electromagnetic emissions that can interfere with other systems that operate in the same location. Some systems are simply noisy or dirty. These characteristics can be harmful to the surrounding environment. Additional design characteristics will be required to limit the environmental harm of any of these issues.

Environment Effect on Operation and Support Personnel

Systems require personnel for operation and support. The environment where the system is used can have a significant effect on personnel. Each of the aspects of environmental impact on personnel must be clearly identified and considered before any possible design solution is developed. Figure 3-7 lists some of the most common environmental issues that must be considered during the acquisition process. The places where the system will be installed or deployed dictate what the environmental issues will be. The technology of the design also will determine any issues caused by the system when used in the specified environment that may be detrimental to anyone who is assigned to operate or maintain the system.

- Geographic locations
- Climatic conditions
- Elements (air, water, etc.)
- Noise
- Space
- Anthropometrics
- Ergonomics

FIGURE 3-7   Environmental effect on personnel.

SYSTEM RATE OF USE

The rate or frequency of use of the system being developed or procured must be determined in order to establish the basis for many subsequent analyses and decisions. There are two statistics that describe the rate or frequency of use: measurement base and the number increments of the measurement base. Figure 3-8 provides examples of several ways that use of a system can be measured, such as time, distance, volume, or events. Automobile usage is measured in the distance, miles or kilometers, driven. Most electronic systems usage is measured in time, usually hours the system is turned on. The usage rate of a pump probably is measured in volume, such as gallons or liters pumped per minute or hour. More complex system may have more than one usage measurement base. For example, usage of an aircraft is measured in several different ways: flying hours for the complete aircraft, operating hours for the engine, and number of landings for the landing gear. A main battle tank also has several measurement bases: operating hours for the engine, distance driven for the track and suspension, and number of rounds fired by the cannon. Once the appropriate measurement base (or bases) has been defined, then the number of increments of each measurement base must be determined. The number of increments combined with the measurement base results in the system usage rate. Figure 3-9 lists examples of system usage rate.

Time
> Years
> Months
> Hours
> Minutes
> Seconds

Distance
> Miles
> Kilometers
> Feet
> Meters

Volume
> Barrels
> Gallons
> Liters
> Cubic inches

Events
> Units produced
> Production strokes
> Transmissions
> Missions per time unit
> Aircraft landings
> Rounds fired

FIGURE 3-8 Usage measurement bases.

Automobile	12,000 miles per year
Computer system	18 hours per day
Transfer press	1,000 strokes per hour
Aircraft	1,200 flying hours per year
	1,500 engine hours per year ·
	600 landings per year
Main battle tank	1,000 miles per year
	2,500 engine hour per year
	100 cannon rounds per year

FIGURE 3-9   System usage rate examples.

Many systems have a consistent usage rate, whereas others have fluctuations in usage patterns. A study of usage patterns can be extremely beneficial. This is a very important point. System usage is an extremely significant statistic that must be determined and agreed on by all parties before any development or procurement decisions are made. Lack of complete definition of system use can result in inappropriate system characteristics, loss or degradation of capability, or overdesign of the system. Any of these unwanted events definitely will result in an undesirable and unnecessary increase in the cost of ownership.

SYSTEM SUPPORT INFRASTRUCTURE

A system must be supported adequately to operate efficiently. Adequate support starts with an infrastructure that can provide resources to operate and maintain the system. Then the resources required to be delivered by the infrastructure must be identified and obtained in sufficient quantities. The actual support resources required to support a system can be identified only after the design solution has been achieved. However, at the start of procurement, it is important to define the current support infrastructure where the new system will operate, assuming that one exists, and resources that may be available when the new system does become operational. This identification is extremely helpful for later analyses, such as standardization and interoperability. Resources that are already available in an existing infrastructure and that can be used again to support a new system tend to lower cost of ownership. Conversely, if the technology baseline of a new system requires support resources that are not currently available, then going to the new technology, while good for the system, may not be as beneficial when considering the overall impact on the support infrastructure.

SYSTEM REQUIREMENTS STUDY

The importance of expending the time and effort to identify the specific requirements for the system cannot be overemphasized. However, the benefits of this activity can be realized only when the requirements are communicated to all organizations and individuals involved in development or procurement of the system. This communication is best achieved through preparation of a single

---

System Requirements Study

1.0	General
1.1	Scope & purpose
1.2	System description
1.3	System mission profile
2.0	Quantitative supportability factors
2.1	Operating requirements
2.2	Number of systems supported and fielding plan
2.3	Transportation factors
2.4	Maintenance factors
2.5	Environmental factors
3.0	Summary of system being replaced
3.1	Operating requirements
3.2	Number of systems supported and locations
3.3	Transportation factors
3.4	Maintenance factors
3.5	Environmental factors
4.0	Support available for new system when in operation
4.1	Maintenance capabilities
4.2	Supply support
4.3	Personnel
4.4	Facilities
4.5	Support equipment
4.6	Test equipment
4.7	Technical data
5.0	Other available supportability information

FIGURE 3-10   System requirements study: document outline.

document that records all the requirements in a logical presentation. Use of such a document to communicate the basis for the new system provides all associated organizations and individuals with a foundation for sound decision making. As procurement of the system progresses, changes to the requirements can be communicated effectively by updating this baseline document. Figure 3-10 provides the standard outline for a system requirements study document. The instructions for preparing the system requirements study can be found in Appendix A.

PERFORMANCE, SUPPORT, AND COST OF OWNERSHIP

The reason for identification of system requirements is to initiate the creation of a range of acceptability that will be used throughout the system acquisition process. The range of acceptability establishes a set of objectives, goals, thresholds, and constraints that are applied to all subsequent activities discussed in this text. The concept of attaining a balance among performance, support, and cost of ownership was introduced in Chapter 1. The parameters defined at the initial stage of a procurement or development program that have been discussed in this chapter form the basis for achieving this balance because the statistics give measurability to system characteristics and to the identification of a reasonable quantity of support resources that will result potentially in the lowest possible cost of ownership.

In subsequent chapters, various analysis techniques will be presented that describe how systems engineering orchestrates the application of supportability engineering principles and analyses in development of ranges of acceptability. The terms *objective, goal, threshold,* and *constraint* will be used to categorize the type of result expected. It is therefore advisable to present a brief description of these four key categories. An *objective* is a clear, unambiguous statement of intent that summarizes the overall effort. A *goal* is something that is the ultimate aim to achieve. A *threshold* is a minimum level that must be attained. And a *constraint* is something that either must be achieved or that cannot be done. Supportability engineering participates in developing the objectives, goals, thresholds, and constraints that will be used to ensure that the final design solution can be supported for a reasonable cost of ownership.

CHAPTER 4
CREATING THE DESIGN SOLUTION

System architecting   The *art* of creating and building complex systems focusing on scoping, structuring, and certification.

System engineering   The *science* of multidisciplinary engineering disciplines in which decisions and designs are based on their effect on the system as a whole.

Design engineering   The *science* transforming a set of functional requirements into a physical entity that possesses a requisite form, fit, and functionality.

The design of a system evolves through a series of engineering activities that start with identification of the functional need for a system and result in the final physical design. Participation in the evolutionary process of creating a design offers significant opportunities for supportability engineering to interject requirements for improving supportability of the design and limiting cost of ownership. This chapter presents an overview of this process.

SYSTEM ARCHITECTING

The first step in creating a system is to identify the limits of acceptability for any system that is ultimately delivered to the user. The role of system architecting in the design process is to establish a creditable range of acceptability within which the design must reside. System architecting is an art, not a scientific process. It is a nonanalytical, inductive approach to creating complex systems. System architecting uses insights, vision, intuitions, judgment, and feelings or taste as primary ways of guiding its activities. It is a proven to be the most appropriate method of creating new and unprecedented systems. One might say that system architecting provides the foundation and framework within which the final system design will be built. Figure 4-1 illustrates some of the reasons why system architecting is applied to the acquisition of a new system.

Most new systems are acquired to replace an aging and possibly obsolete system. The old system may have been in use for many years, and a tremendous amount of usage and support data are available for analysis. However, the old system's technology baseline also may be obsolete, the environment of use has altered, and the intended use of the new system may be broadened or changed significantly. All these situations make the historical data on the old system virtually useless in development of the new system.

Any attempt to perform a detailed "scientific" analysis when there are no boundaries or limits can be simply overwhelming. To be effective, a scientific analysis must investigate facts. At the beginning of development of a new system, there are just too many unknowns. Assumptions must be made. Decisions about the unknowns of the future must be defined using some other method. This is the purpose of system architecting—to establish some definition of limits for the new system.

The start of a program needs direction. Imagine a person who sets out on a journey. With no direction, the person will wander aimlessly with no ultimate goal to complete the journey. There must be

- Past system data of limited use
- Scientific analysis overwhelmed by too many unknowns
- Too many possibilities
- Too little time for data gathering and analysis

FIGURE 4-1   Reasons for system architecting.

some initial definition of goals to be reached. The same is true for an acquisition project. There are too many possibilities without some clear understanding of the final goals coupled with limitations and boundaries.

Finally, the design of a system is time-dependent. It is created by a fast and furious process. There is normally too little time for gathering relative data and then analyzing them. The design processes needs answers immediately, not sometime in the future.

Unmeasurables

There any many issues that the design process must consider. It would be nice if each issue could be measured and analyzed scientifically. But many of the most important issues that guide the initial development of a system are unmeasurable. Figure 4-2 lists some of the most common unmeasurable areas addressed by system architecting. Each of these issues may pose a significant concern that must be considered at the very initial stages of system acquisition. Therefore, they all must be considered at the start of an acquisition.

Public opinion must be considered. If the public has concerns about the viability of a system, the concerns must be addressed. There are many ideas that might be applicable to a system design, but public opinion against an idea would doom it for failure. If the system is to be used by or provide services to the public, then it must be acceptable. Market surveys, opinion polls, and consumer research provide valuable insight into identification and resolution of these issues.

There may be systems that required political backing or funding. A system characteristic or technology that is not politically acceptable should be avoided. This has been a consistent issue with systems with military applications. This issue must be combined with the view of public opinion because political acceptance typically is swayed according to considerations of public acceptability.

Any final design solution must conform to some perceived environmental acceptance. The question that must be answered is the effect that the new system will have on the environment in terms of

- Political acceptance
- Environmental impact
- Public opinion
- Perceptions or reality
- Safety
- Security
- Affordability
- Worth or value for money

FIGURE 4-2   System architecting interpretation areas.

resources obtained from the environment and by-products or emissions that will be placed into the environment. There are acceptable limits for both these concerns. For example, a system that produces a noxious discharge would not be acceptable in a populated area, or a system that produces ozone-depleting emissions would not be acceptable. If a system requires a significant quantity of fuel to operate, the sources of the fuel must be assessed.

Another challenge that system architecting must deal with is the difference between perceptions and reality of the user. The user may have a perception of what the requirement is and what a system must be to meet the need. However, that perception may be very different from reality. There is a famous quote from an unnamed politician saying, "Truth is what people believe, and that is not necessarily the real truth." System architecting must separate perception from reality and only incorporate reality into its activities. Then it must "sell" the reality to change perception.

Safety

There is always a concern about the safety of a system. It is very important that a system can be operated and maintained as safely as possible. The safety potential of a system sometimes may be dictated by its technology base, environment of use, and usage profile. System architecting must consider these three issues when developing the range of acceptability of any aspect of the design. The inherent capability of an item to be safe to operate and use many times depends on the capabilities and limitations of the user. A car is inherently safe. It only becomes potentially unsafe when in operation. Therefore, system architecting must consider the user's interaction with the system as a primary area for development of a system that optimizes safety.

Some systems have an inherent requirement for security. For example, data systems must have security features to protect their contents from unauthorized access or modification. Many military systems must consider national security implications in their design. The need for security may create an additional minimum feature for a system being designed.

Affordability

The initial foundation for the design of a new system that is established by system architecting must consider affordability, not cost of ownership. Affordability looks at the larger scale to determine the range of expenditure of money that might be caused by any potential design. For example, a person in need of transportation does not immediately go purchase a car. The wise person first considers the affordability of any transportation means. Can the person even afford a car? It may be more appropriate to consider alternate means of transportation, such as bus, train, etc. The same is true for system architecting on a new design project.

In addition to affordability, system architecting must consider the worth or value for money that the user will place on the system. The user must have a sense of getting value from the expenditure of money and other resources to possess the system. The higher the value placed on the system, the more consideration must be given to performance and supportability characteristics with cost of ownership.

None of these issues can be measured, but all form a significant set of concerns that must guide initial establishment of the basic parameters that will be used to create the system. It is the responsibility of system architecting to produce a foundation for the new system that conforms to these parameters. A system that does not conform to the limitations of these issues will not be acceptable regardless of its performance capabilities.

System Architecting Methodologies

There are four methods that can by applied by system architecting to define system parameters. These methods, listed at Figure 4-3, can be both qualitative and quantitative. When used together, these methods provide the basis for determining the range of acceptability for any design solution.

- Normative (working within predetermined solution parameters)
- Rational (analytical techniques)
- Consensus (group solutions)
- Heuristics (past experience and lessons learned)

FIGURE 4-3   System architecting methodologies.

The *normative method* of system architecting identifies established parameters for the type of system being developed. These parameters may be prescribed by established codes, such as building codes of a city, or regulatory agencies. Authoritative organizations or associations may have published standards to which the system must conform such as Institute of Electrical and Electronics Engineers (IEEE) standards, International Standards Organization (ISO) specifications, military standards, and trade handbooks. System architecting identifies the applicable established parameters and then expresses them in terms appropriate for the new system.

The next step of system architecting is to apply *analytical techniques* to rationalize parameters for the system. This analysis is performed for system-level characteristics to determine any measurable points for the parameters. Conceptual modeling using appropriate assumptions and usage projections is beneficial for this step. The modeling result can be used to determine limits, both minimum and maximum, for the system that point toward achievement of necessary levels of operational effectiveness (O_E), operational availability (A_O), and cost of ownership. This modeling is to determine what the system needs to attain, not what it actually will achieve.

The third step of the system architecting process is to obtain input from all the groups participating in and associated with the system with a view toward *consensus*. This group includes the user, all engineering specialties, manufacturing, quality, finance, and especially supportability engineering. System architecting takes the inputs from each group and attempts to develop a set of system parameters that each group agrees will meet their specific concerns. The focus is to further define the qualitative and quantitative parameters so that the final design foundation will produce a system that each group will accept as a reasonable solution.

The final step of this process is to apply common sense. The formal name for this method is *application of heuristics*. A *heuristic* is a lesson, or truth, that has been learned in a previous specific situation that is proven to be applicable in a general sense to all future situations. Figure 4-4 lists typical heuristics. Each of these is common sense and self-explanatory. However, they apply to any system development project. Probably the most applicable is the challenge for simplicity, and then always looking for ways things can go wrong. It is always wise to research the basic premises on which the

- Keep it simple (the KISS principle).
- If it can go wrong, it will (Murphy's law).
- Don't assume that the original statement of the problem is the best or even the right one.
- If you can't explain it in 5 minutes, either you don't understand it or it doesn't work.
- All the serious mistakes are made on the first day of a project.
- A model is not reality.

FIGURE 4-4   Heuristics examples.

project is established because the original problem statement may not be the real problem. Anytime that modeling techniques are applied to a project, especially in its formulative stages, everyone must understand that the model results are not true, only projections of possibilities.

Complexity

System architecting may be a relatively new topic to some readers. It has evolved with the advent of increasingly complex systems. Figure 4-5 shows how complexity of systems grows as functions are added. As each additional function is integrated into the system, its complexity increases exponentially. The challenge for system architecting is to understand the interrelationships of all the system functions and produce a set of parameters that guide all design and development activities. Supportability is an integral part of this activity. Chapter 6 presents a detailed discussion of supportability characteristics that must be included in the system architecting process.

System Architecting Results

The system architecting process results in definition of the parameters to which the system design must adhere. The parameters are best described as issues and considerations that are expressed in quantitative terms, where possible, limitations or boundaries. They form the goals, thresholds, and constraints for the systems engineering process. Figure 4-6 illustrates this concept. Parameters are identified as mandatory, desirable, undesirable, and unacceptable. A *mandatory* parameter must be achieved. An example might be that an aircraft must be capable of reaching a speed of Mach 2.2. In this example, the word *must* makes it a mandatory requirement. A *desirable* parameter is one that should be achieved, if possible. An *undesirable* parameter is one that should not occur but may be acceptable if no other alternative is possible. An *unacceptable* parameter is an outcome that will not be accepted. An example of this might be that the design will require no scheduled maintenance. Here, the word *no* declares that any scheduled maintenance would be unacceptable.

The system architecting process provides a clear definition of the range of acceptability for performance, supportability, and cost-of-ownership parameters for the new system. These parameters

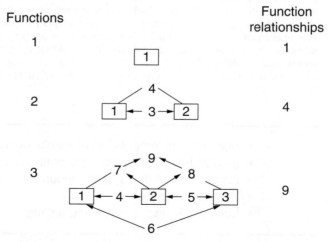

FIGURE 4-5 Design functional complexity growth.

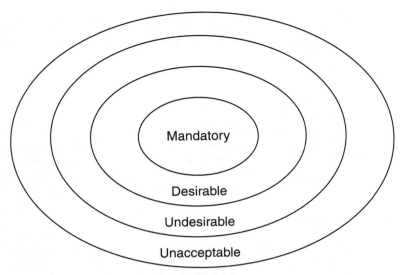

FIGURE 4-6   Establishing a range of acceptable system parameters.

then are passed to the systems engineering process for scientific implementation in the final design solution.

SYSTEM ENGINEERING

The system engineering process applies scientific analysis and engineering techniques and methods to create a system that meets a specified need. The method for identification of the need was presented in Chapter 3. This definition of the need gives the primary direction for all system engineering activities. Figure 4-7 lists the five responsibilities of system engineering.

The range of acceptable system parameters produced by system architecting is the basis for all system engineering activities. Each of the parameters is analyzed to determine the most appropriate method of describing its measurable characteristics. Where a specific parameter cannot be translated into a measurable requirement, system engineering investigates alternatives for expressing the requirement or investigates subelements of the parameter that can be measured. The measurability of

- Scientific implementation of results of system architecting
- Establishment of measurable characteristics
- Preparation of design and procurement instructions
- System integration
- Measurable assessment and testing

FIGURE 4-7   System engineering responsibilities.

each parameter is mandatory. If a requirement cannot be measured, then its achievement cannot be verified.

System engineering then is responsible to ensure that instructions issued to design engineering for creation of the physical items within a system contain the appropriate measurable parameters. Measurable parameters also must be issued to procurement activities that outsource portions of the design. This ensures that parameters are "built in" to the evolving design solution.

As each lower-level item design is completed, system engineering is responsible for its integration into the overall system solution. This is accomplished by definition of the interfaces between each item within the total system. Integration is a bottom-up construction of the physical system architecture.

The final system engineering responsibility is assessment and testing of the final design solution to verify that all measurable requirements have been achieved. Assessment and testing can be performed by analysis, demonstration, and testing. System engineering determines the most appropriate method for verifying achievement of each parameter.

System Engineering Process

There are many excellent textbooks devoted to the system engineering process. It is suggested that readers refer to one of these for a detailed description of system engineering. The purpose of this section is to highlight how supportability engineering functions as an active participant in the system engineering process. Every decision made during the system engineering process must consider the effect of that decision on performance, supportability, and cost of ownership. The areas of supportability and cost of ownership are the focus of supportability engineering.

The development cycle of a system is illustrated at Figure 4-8. The system starts with the definition of the need. Chapter 3 presented the method for defining the need, including mission, measurement basis, environment of use, and all other pertinent issues that must be considered for any design solution. Once the need has been defined completely, then the project moves into the conceptual design stage. This stage starts with system architecting and then moves to system engineering. The next stage is development of a preliminary design solution. Then the detailed design activity produces the product baseline that is passed to manufacturing for actual production of the system. The final system is delivered to the user, where it is placed into service and is sustained throughout is useful service life. System engineering is a team effort with participation of all development disciplines. Figure 4-9 lists the typical composition of the system engineering team. Supportability engineering participates in all phases of system engineering as the system progresses through each stage.

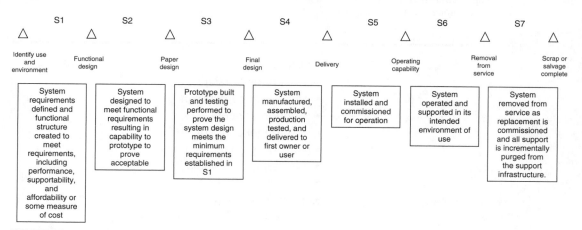

FIGURE 4-8   System life cycle.

- Systems engineer
- Electronic engineer
- Mechanical engineer
- Software engineer
- Human factors engineer
- Safety engineer
- Production engineer
- Supportability engineer
- Reliability engineer
- Maintainability engineer
- Testability engineer
- Quality engineer

FIGURE 4-9   System engineering team.

Conceptual Design Stage

Supportability engineering focus during the conceptual design stage centers on development of requirements that eventually will produce a supportable design. These requirements are stated in terms of objectives, goals, thresholds, and constraints. Each requirement must be determined in relationship to the overall project goals of availability and cost of ownership.

Targets for inherent availability, achieved availability, and operational availability must be established. These are a combination of all other individual targets. This is mandatory to have the ability to analyze various design and technology alternatives that are investigated by system engineering during this stage.

System engineering typically develops several design approaches to meet the stated user need. Each design approach is assessed by supportability engineering to determine the preferred maintenance, support, and personnel concepts for each design approach. Figure 4-10 indicates the areas that supportability engineering addresses during the conceptual design stage. This is an aid to system engineering in evaluating the overall goodness of each design alternative in an effort to achieve a balance among performance, supportability, and cost of ownership.

- Availability
- Cost of ownership
- Maintenance concept
- Support concept
- Personnel concept
- Reliability goals
- Maintainability goals
- Testability goals
- Standardization
- Interoperability

FIGURE 4-10   Supportability in the conceptual design.

Reliability engineering, maintainability engineering, and testability engineering develop specific targets for their area of expertise. Supportability engineering assists system engineering in determining the reasonable balance between these targets and the overall system availability targets. This is especially important for the system operational availability target.

Finally, supportability engineering identifies potential standardization and interoperability issues for inclusion in the system design requirements. Standardization tends to have little effect on performance, but it can have significant benefits in terms of cost of manufacture and cost of ownership.

The concept of cost of ownership is presented in Chapter 10. An important fact discussed there is that studies show that 70 percent of the cost-of-ownership decisions made on an acquisition project are made during the conceptual design stage. This is why participation of supportability engineering is so important. Specific design characteristics that improve supportability are discussed in Chapter 6. The methods used by supportability engineering during the conceptual design stage are discussed further in Chapter 11.

Preliminary Design Stage

One of the important tools used by system engineering in the preliminary design stage is development of a system functional block diagram. This diagram serves as the basis for determination of the functions required for the system to meet its operational need. Figure 4-11 is an example of a system functional block diagram. The block diagram starts with the top-level system functions and then divides the functions into subsystems, assemblies, and finally, the lowest designable or procurable item. This is the basis for development of the final system architecture. Chapter 5 discusses how this diagram is used by reliability engineering to prepare reliability goals and then predictions of system reliability. It is also used by maintainability engineering in a similar way for maintainability goals and then prediction of system maintainability. These determinations form the basis for projection of system availability.

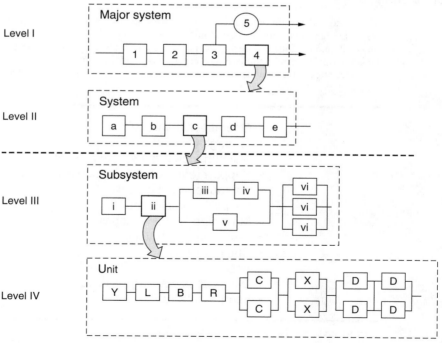

FIGURE 4-11   System functional block diagram.

- Availability
- Life cycle cost
- Maintenance planning
- Support resources
- Personnel utilization
- Reliability allocations
- Maintainability allocations
- Testability allocations
- Standardization
- Accessibility
- Diagnostics
- Interoperability

FIGURE 4-12   Supportability in the preliminary design stage.

Supportability engineering activities during the preliminary design stage are a refinement, amplification, and expansion of the activities started during the conceptual design stage. Figure 4-12 lists many of the significant areas that supportability engineering addresses. It is notable that supportability engineering efforts during this stage differ from efforts during the conceptual stage with regard to each of these areas. Conceptual design efforts focus on what is required; however, during preliminary design, supportability engineering efforts focus on how requirements can be achieved.

Detailed Design Stage

The functional system requirements are transformed to a physical entity during the detailed design stage. Design engineers are issued instructions for the form, fit, and functionality of an item. The physical item produced by design engineers must possess the proper supportability characteristics to be supportable. Supportability engineering works with design engineers in implementation of these characteristics in the final design solution. Figure 4-13 shows the important areas that supportability

- Operational availability
- Through-life cost
- Maintenance requirements identification
- Support resource requirements
- Personnel requirements
- Reliability predictions
- Maintainability predictions
- Testability predictions
- Standardization
- Accessibility
- Diagnostics coverage and accuracy
- Interoperability

FIGURE 4-13   Supportability in the detailed design stage.

engineering continually analyzes as the design becomes a reality. This implementation stage should produce a system that attains a reasonable balance among performance, supportability, and cost of ownership. The individual design engineer does not have visibility as to how his or her piece of the system fits into the overall big picture. It is supportability engineering's responsibility to provide this visibility.

Design Prototype and Testing

The paper design produced by design engineers must be tested to ensure that it meets the performance requirements delineated in the specification. This is accomplished by building a prototype system that is subjected to sufficient testing methods to ensure that the design meets its requirements. The difference between the prototype and a standard production system varies. The prototype may be exactly like the standard production model except that the prototype was build using "model shop" techniques rather than standard manufacturing processes. There may be other differences, such as lack of proper finishes or color variance. The critical issue is that the prototype must strictly meet the form, fit, and function parameters that will be continued in the standard production models. Supportability characteristics must be present in the prototype and must be tested along with performance characteristics. A detailed discussion of supportability testing and assessment is provided in Chapter 13.

Production and Delivery

Supportability engineering activities for production and delivery should be focused on delivery of a reasonable support-resource package. Figure 4-14 indicates typical activities during this stage. There should be relatively minimal design activities, assuming that the final system design achieved the required supportability characteristics. However, there can be design changes necessary for producibility improvement. Each design change must be assessed to determine any effect on supportability or the support-resource package. Where a design change affects either supportability or the support-resource package, supportability engineering must ensure that appropriate actions are taken to limit or avoid the impact of the change.

In-Service Sustainment

Introduction of the system into service presents a different set of challenges for supportability engineering. This stage of the development cycle has two significant activities: assessing the achievement of supportability and formulation of modifications and upgrades to either resolve shortfalls in the original design or modifications and improvements that increase the supportability potential of

- Resource production and delivery
- Prediction update
- Through life cost
- Establish support capability
- Assess design changes
- Plan post delivery assessment

FIGURE 4-14   Supportability in the production and delivery stage.

- Post delivery assessment
- In-service data collection
- Trend analysis
- Design change assessment
- Modification planning

FIGURE 4-15   Supportability in the in-service stage.

the system. The basis of activity comes from the postdelivery assessment. This assessment is performed by the user. It provides specific feedback on the supportability characteristics of the design and the adequacy of the support-resource package that was delivered with the system. Supportability engineering continually assesses, improves, assesses, and improves the system design and support infrastructure throughout the life of the system. This activity also constitutes the start of planning for the system to replace the in-service system in the future (Figure 4-15).

CHAPTER 5
RELIABILITY, MAINTAINABILITY, AND TESTABILITY

Supportability engineering activities encompass all aspects of system design. A system that does not have adequate and appropriate supportability characteristics cannot be supported. Three of the principal supportability characteristics of any design are its reliability, maintainability, and testability. These three areas contribute significantly to the overall supportability of the system. Failure modes, effects, and criticality analysis (FMECA) is one of the most important design tools available to integrate these engineering disciplines into the systems engineering process. Therefore, a detailed discussion of each of these engineering specialty disciplines is important to understand how they participate in design decisions, the various analysis techniques used, and how their efforts ultimately lead to a supportable design.

RELIABILITY ENGINEERING

Reliability is the term used most commonly to express a desire or confidence in a system to perform as required, but sooner or later all systems break. The more often a system breaks, the less amount of time it is available for use, and the greater is the amount of resources required to support it. The purpose of reliability engineering is to address this problem of how to limit the number of times that an item will break over its useful life. Reliability engineering's effort is really twofold: (1) participation in the design and development of the system to make it as failure-free as possible, and (2) predicting how the system will fail when it is being used so that adequate resources will be available to make repairs. The results of these tasks are used by supportability engineering to develop design improvements, estimate the cost of ownership, and support resource requirements.

Before starting a lengthy discussion of reliability engineering, the following definitions are necessary:

Reliability engineering   Application of a standard set of mathematical or statistical methods and analyses to predict the reliability of an item and to identify where reliability of an item can be improved by design changes.

Reliability   The probability that a system will perform its intended mission without failing, assuming that the item is used within the conditions for which it was designed. Note that there are two key statements in this definition: (1) the word *probability* indicates that reliability deals with statistical calculations and projections of how and when failures may occur, and (2) *within the conditions for which it was designed* sets a boundary on the validity of any predicted or perceived level of reliability.

Failure   Any deviation from the design-specified, measurable tolerance limits that causes either a loss of function or reduced capability.

Failure rate (λ_I)   The predicted number of inherent failures during some period of system operation.

Mean time between failures (MTBF)   The predicted elapsed time between inherent failures of a system during operation.

Mission failure rate (λ_M)   A prediction of mission failures estimated to occur during some period of system operation.

Mean time between critical (mission) failures (MTBCF)   The predicted elapsed time between mission failures of a system during operation.

Mission reliability   The probability that a system will not experience a complete failure, or loss of all function, during the performance of a specified mission or use.

Fault   The physical or chemical processes, design defects, quality defects, part misapplications, or other processes that are the basic reason for failure or that initiate the physical process by which deterioration proceeds to failure.

Failure mode   The manner by which a failure is observed.

Failure effect   The consequence(s) a failure mode has on the operations, function, or status of an item.

Concept of Reliability

The reliability of an item is contingent on numerous factors. Figure 5-1 illustrates the basis for the reliability of an item. Note that a stated level of reliability is a probability, a statistic, not reality. Reliability is used to portray a sense of goodness or confidence in an item. In order to start considering the reliability potential of an item, the conditions and situation under which the item will be used must be defined. Chapter 3 described how the parameters of use for a system are developed as an initial step in development or procurement. These parameters are the basis for determining the reliability of an item. Stated another way, once the parameters of use for a system have been specified, it is then possible to predict the reliability potential of an item to operate within those parameters. Reliability is based on a prespecified set of circumstances. If a system is used outside these specified parameters, its reliability most probably will be different. For example, if an item that has been designed to operate within a temperature range of +5 to +35°C and is actually used in a place that has a temperature range of −20 to +65°C, it probably will fail at a higher rate because the temperature range is significantly more harsh and places more stress on the item. Therefore, the parameters of use are extremely important in terms of reliability potential of an item.

Reliability Statistics

The reliability of a system is expressed in terms of the probability that it will perform its intended functions successfully. Therefore, all reliability is based on statistics. Any statistic must originate from a known source. Typically, reliability statistics developed for a system come from one of the three sources listed at Figure 5-2.

The actual in-service experience data from a similar system can be very useful in predicting the reliability potential of a new system. There are some limitations on this type of data source. The environment of use and the use profile must be very near those of the new system or the data from the experience of the similar system may not be relevant.

The *probability* that an item will

    a. Perform its *specified function(s)*
    b. For a *specified interval*
    c. Under *specified conditions*

FIGURE 5-1   Concept of reliability.

- Use of in-service data from similar equipment
- Extrapolation of test or trials data
- Generic parts data

FIGURE 5-2 Reliability statistics sources.

When in-service experience data is not available or not relevant, then another source may be test data from prototypes of the new system. Typically, new systems progress through a series of development tests for proof of concept and functionality. This data source may be much more realistic than data from a similar system. There are significant limitations on the reality of test data, however. Normally, the test circumstances do not reflect the actual usage rate or environment for the new system, so the data resulting from testing may prove to be invalid. However, this data source can be useful if no other source is available.

The third method of developing reliability statistics uses data on the components or pieces that make up the new system. Techniques presented later in this chapter will show how such data is used to construct a prediction of new system reliability. This method can be useful where no similar system data is relevant and no testing results are available.

Reliability statistics, regardless of the data source, attempt to predict the probability of system reliability in the future within the specified conditions of use and environment. These statistics can be either positive, potential for success or negative, potential for failure. Figure 5-3 shows that reliability can be the basis for predicting that the system will perform its mission successfully without failing, or the statistics can predict the frequency that items within the system will fail or the frequency that the total system will fail to perform its mission. These different views of system reliability are interrelated, but each provides a specific set of knowledge about the system.

Failure Rate (λ_i)

The failure rate for a system or an item within a system indicates the anticipated frequency that a failure will occur. The failure rate is a numeric value that predicts the number of failures of a system, item, assembly, or piece part that will occur during a specified period of operation. Failure rates are developed using the sources identified earlier. Figure 5-4 illustrates how a failure rate can be calculated using actual usage data. The number of failures that occurred over a specific length of time, when divided by the length of time, results in the failure rate. The resulting failure rate can be applied to any length of operation to determine the predicted number of failures that may occur. Note that the

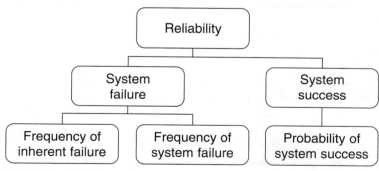

FIGURE 5-3 Reliability statistics.

$$\lambda_I = \frac{\text{Number of failures}}{\text{Total measured usage}}$$

Example: A system experiences 4 failures during 500 hours of operation

$$\lambda_I = \frac{4 \ \text{(failures)}}{500 \ \text{(operating hours)}} = 0.008 \ \text{(failures per operating hour)}$$

FIGURE 5-4   Failure rate calculation formula.

Greek letter lambda (λ) is used to represent the failure rate. As shown in Figure 5-4, failure rates normally are numbers with several decimal places. A common standard method for writing failure rates is failures per million hours of operation. For example, one failure experienced in 1 million hours of operation (1/1,000,000 or 0.000001) would be written 1.0×10^{-6}. The failure rate calculated at Figure 5-4 would be written as 8000×10^{-6}.

A failure rate can be calculated for any item, assembly, or part within a system. Typically, each item in the structure of a system has its own failure rate. These individual failure rates then are summed to calculate the failure rate of the system. Figure 5-5 shows how this is done. The failure rates of each item within the system are based on the same principle as that for the system. Having individual failure rates for each item within a system allows identification of the items that are driving the overall failure rate of the system. The example in Figure 5-5 shows that assembly B contributes 50 percent to the system failure rate. It therefore becomes the first target for improvement of system reliability. This concept will be investigated later in this chapter.

Mean Time between Failures (MTBF)

Another expression of reliability is the mean time between failures (MTBF). Calculation of an MTBF is accomplished using the same data as used for calculating a failure rate. Figure 5-6 shows how an MTBF is calculated. Since both the MTBF and the failure rate are calculated using identical data, they are the same statistic presented in two different ways. The failure rate that predicts the average number of hours that an item, assembly, or piece part will operate before it fails, and the MTBF predicts

$$\text{System} \ \lambda_I = \sum_{I}^{n} \lambda_I$$

Assembly *A* $\lambda_I = 0.001$	Assembly *B* $\lambda_I = 0.004$	Assembly *C* $\lambda_I = 0.003$

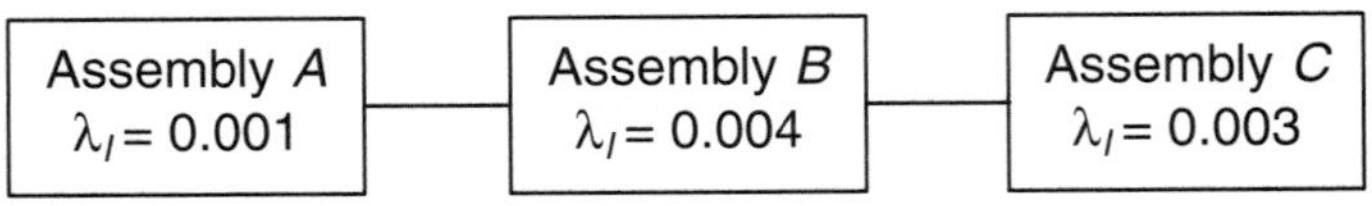

$$\text{System} \ \lambda_I = \lambda_A + \lambda_B + \lambda_C = 0.001 + 0.004 + 0.003 = 0.008$$

FIGURE 5-5   Calculating system failure rate.

$$MTBF = \frac{\text{total measured usage}}{\text{number of failures}}$$

$$MTBF = \frac{500 \text{ (operating hours)}}{4 \text{ (failures)}} = 125 \text{ hours}$$

FIGURE 5-6   Mean time between failures.

the average elapsed time between failures. Figure 5-7 shows that the MTBF is the reciprocal of the failure rate, and the failure rate is the reciprocal of the MTBF. An item with a failure rate of 0.008 would have an MTBF of 125 hours. The MTBF of an item frequently is used in calculating spares requirements, life-cycle cost, and other failure-related data. Practice has shown that it is easier to use an MTBF when talking about the reliability of an item than a failure rate because people tend to be more comfortable with whole numbers rather than decimals.

System Mission Failure Rate

A system consists of items that have been assembled to perform a specific function. Each item within the system is related to its overall capability. When one item fails, the system fails. This can be compared with a chain, where each link must be is of equal value in the functionality of the chain. When a link fails, the chain fails. Figure 5-5 illustrates a system where the assemblies are linked in series to achieve system functionality. If any of the assemblies fail, the system can no longer perform its required functions. Therefore, as illustrated in Figure 5-5, the sum of the individual failure rates is equal the failure rate of the system. This is true when any item failure causes the system to fail. However, some systems, because of criticality of a function, are designed to be able to experience an item failure but continue to perform the required function. This design feature is called *parallel redundancy.*

One of the predictions that is used to gauge how an item might perform is its predicted mission reliability. This is expressed as the probability of completing a mission of a specified length successfully. Figure 5-3 provides the formula for this prediction. Although the results of this calculation may not look like it has any usable relationship to the actual performance of the item being developed, it does provide a baseline for evaluation of tradeoffs and design changes that affect reliability.

Redundancy   A design characteristic where two or more like functions are incorporated into a system so that when an inherent failure occurs, the system is still mission-capable.

$$MTBF = \frac{1}{\lambda_l} \quad \text{or} \quad \lambda_l = \frac{1}{MTBF}$$

$$MTBF = \frac{1}{0.008} = 125 \text{ hours}$$

FIGURE 5-7   Failure rate and MTBF relationship.

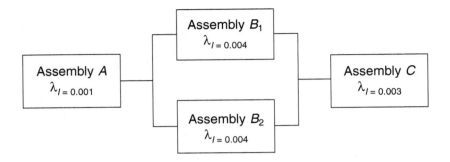

$$\text{System } \lambda_I = \lambda_A + \lambda_{B_1} + \lambda_{B_2} + \lambda_C = 0.001 + 0.004 + 0.004 + 0.003 = 0.012$$

FIGURE 5-8   System inherent failure rate.

When active parallel redundant items are incorporated into the design of a system, there is a divergence between the sum of the inherent failure rates of individual items and the mission failure rate of the system. This creates some confusion when discussing the reliability potential of an item. Figure 5-8 shows that the series linkage illustrated in Figure 5-5 has been modified by adding an additional assembly B in parallel redundancy. The addition of the second assembly has increased the inherent failure rate of the system; however, the redundancy must be considered when calculating the rate at which the system will fail, which would occur only if both assembly B_1 and assembly B_2 failed. Therefore, we have the basis for having two failure rates for the system. One failure rate (λ_I) will predict the inherent failure rate of the system, and the other (λ_M) will predict the mission failure rate of the system. Figure 5-8 shows how the mission failure rate (λ_M) is calculated by incorporating the joint failure rate of the parallel assemblies into the system mission failure rate. This calculation is valid only where items are in active parallel redundancy.

Mean Time between Critical (Mission) Failure

The system mission failure rate is a valuable statistic in determining the frequency at which the system will experience a mission failure. As was discussed previously, a number with several decimal places tends to be cumbersome to use, so the reciprocal of the system mission failure rate translates this number into a mean time between critical or mission failures (MTBCF). As shown at Figure 5-9, the incorporation of a parallel redundant assembly B produces an MTBCF of 166.67 hours; however, it has been done with the penalty of reducing the system MTBF to 83.33 hours. The addition of a second assembly B in the system architecture also will result in a higher system cost, which is another penalty.

The difference between system inherent failure rate and system mission failure rate, or system MTBF and system MTBCF, is a significant issue. These two figures of merit have several applications that will be discussed in detail in later chapters in this book. In general terms, the user is most concerned with MTBCF, which indicates that the system potentially will operate longer without a mission failure. However, the supportability engineer also must consider that to achieve a higher MTBCF, the system MTBF has been reduced, thereby increasing the frequency of requirements for maintenance. Hence the system acquisition cost has increased, and the cost of support through life has increased in order to increase the user's confidence in the system's mission reliability.

This is a prime example of the ultimate tradeoff issues facing the organization making development and procurement decisions—assessing every possibility and their relationships in order to

$$\text{MTBCF} = 2/(\lambda_B + \lambda_A + \lambda_C) - 1/(2\lambda_B + \lambda_A + \lambda_C)$$

$$\lambda_B = 0.004 \text{ fph}$$

$$\lambda_{A+C} = 0.001 + 0.003 = 0.004 \text{ fph}$$

$$\text{MTBCF} = 2/(0.004 + 0.004) - 1/[(2 * 0.004) + 0.004]$$

$$\text{MTBCF} = 250 - 83.33 = 167.667 \text{ hours}$$

$$\text{MTBF} = \frac{1}{\text{system } \lambda_I} = \frac{1}{0.012} = 83.33 \text{ hours}$$

FIGURE 5-9   Mean time between critical failures.

arrive at a final decision concerning the system. There is no magic answer, only the most reasonable balance between alternatives when all issues have been considered, achieving a reasonable balance among performance, support, and cost.

Mission Success

The third reliability statistic addresses the probability that the system will complete its mission without experiencing a complete failure. This statistic can be computed using various probability distribution methods, such as exponential, binomial, normal, Poisson, gamma, and Weibull. The most common method used is based on exponential distribution using the formula shown in Figure 5-10. This formula can be used to calculate the probability that an item will not fail for a specified use measurement base, such as hours, miles, rounds, etc. The result of this formula produces a confidence in the probability that the system will not fail when used under the specified conditions and environments. The obvious goal for a system is 1.00, or 100 percent, but nothing in life is perfect. Thus the typical acceptable result is in the range of 0.95 (95 percent) to 0.99 (99 percent).

Applying the exponential formula to the individual failure rates presented in Figure 5-5 produces the results shown in Figure 5-11. These individual reliability probabilities then are multiplied to produce a probability of mission success. In this example, the resulting calculations indicate that there is a 93.8 percent probability that the system will complete a mission of 8 hours without failing.

These three statistics—frequency of item failure, frequency of mission failure, and probability of mission success—are the foundation for further supportability analyses to identify the significant

$$R(t) = e^{-\lambda t}$$

where $R(t)$ = Reliability that system will not fail to time (t)
    e = Natural logarithm base (2.7183)
    λ = Inherent failure rate of the item
    t = Time of the mission or use

FIGURE 5-10   Exponential distribution for mission success.

Assuming the mission time (t) is 8 hours:

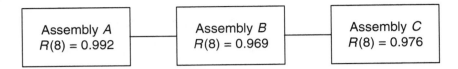

Assembly A R(8) = 0.992	Assembly B R(8) = 0.969	Assembly C R(8) = 0.976

Formula for system reliability (R_S):

$$R_S = (R_A)(R_B)(R_C)$$

Therefore:

$$R_S = 0.992 \times 0.969 \times 0.976 = 0.938$$

Result: There is a 93.8% probability that the system will

complete an 8-hour mission without failing

FIGURE 5-11   Probability of mission success.

support issues and to establish targets for improvement. Reliability engineering is a critical ingredient for system success and should be applied on every project.

THE RELIABILITY PROGRAM

Reliability engineering approaches its responsibility for participation in a development or procurement project through application of a series of commonsense-based activities consisting of management, analyses, and testing. These activities provide a consistent methodology for predicting the reliability potential of a system, testing the system to verify the achievement of a degree of reliability, and managing the reliability engineering process.

Reliability Design and Evaluation

Reliability engineering activities for participation in the development or selection of a system design and then evaluation of the resulting configuration may consist of up to nine discrete but interrelated analyses. These analyses, listed in Figure 5-12, are applied as appropriate to specific projects; however, the first four on the list should be applied to every project. Accurate and successful performance of these tasks is the key to developing a reliable item for a system.

Reliability Modeling

As stated previously, reliability is a statistical study. Therefore, the first step in initiating the statistical analysis is to develop a creditable method for modeling the reliability of an item. Reliability

- Reliability modeling
- Reliability allocation
- Reliability prediction
- Failure modes effects and criticality analysis
- Sneak circuit analysis
- Tolerance analysis
- Parts control program
- Reliability critical items analysis
- Effects of testing, storage, packaging, handling, transportation, and maintenance

FIGURE 5-12 Reliability design and evaluation analyses.

mathematical models are developed based on the functional operations of the system being analyzed and are tailored to reflect the specific operating requirements of the system. The purpose of modeling is to provide an accurate method of predicting failure rates and probability of success. In the early stages of a system development program, reliability engineering develops a reliability block diagram. The purpose of the block diagram is to provide a base for accurate mathematical model preparation. The block diagram is constructed to represent the functional configuration of the system. Normally, reliability engineering uses the same functional block diagram that was developed by systems engineering, as discussed in Chapters 3 and 4. This ensures compatibility between analyses performed by systems engineering and those performed by reliability engineering. The functional block diagram then is the basis to develop a mathematical model that reflects the interrelationships between functions that eventually will become hardware and software.

After the reliability block diagram has been constructed, reliability engineering moves to selection of the most appropriate method for projecting the reliability statistics for the new system. Previously in this chapter, the techniques for estimation of item and system failure rates were discussed. These techniques are used to develop the reliability potential of a system. This is a step-by-step process. The first step is to create a reliability target for the system. On a project for development of a new system, it is vital that a reliability target be established. This becomes the reliability "design to" goal for the project. This goes back to the issues discussed in Chapters 1, 2, and 3, where the need for the system is identified. For example, a communications company initiates a project to develop a communications satellite. The reliability of the satellite is critical to the company's business success; therefore, it must be designed to operate without mission failure. Assuming that the satellite will not be maintained after launch into orbit, it must be designed to operate successfully for its useful life. In this example, reliability engineering must establish a system mission success reliability goal, or target. Figure 5-13 illustrates how the techniques discussed previously in this chapter can be used to establish this goal. As noted in the example, the result is the system mission failure rate and MTBCF. This reliability figure of merit then is used as a basis for all design and procurement activities.

The reliability goal is the desired result of the reliability functional block diagram. The next step is to select the most appropriate method to project the reliability requirements to meet the goal. The types of methods available to use on a project were listed previously in this chapter, with exponential distribution being identified as the method selected most often. The method selected then is applied to the functional block diagram to identify requirements for redundancy, critical reliability items, and the participation of each item, assembly, and subassembly in achieving the system goal.

Situation:	Satellite must operate for 10 years Design must achieve a 99% probability of success
Formula:	$R(t) = e^{-\lambda t}$
Method:	$R(t)$ must result in 99% t = 10 years = 8760 hours per year $\times$ 10 = 87600 hours Solve equation for λ_M (in this case, mission failure rate)
Results:	To achieve 99% probability of mission success, the satellite system mission failure rate goal is 0.00000011473, which is an MTBCF of 8,716,116 hours

FIGURE 5-13 Establishing a reliability goal.

Reliability Allocation

During the early design phase of a system, it is important to have a reliability baseline to guide design engineers as the system is developed. The reliability goal established through modeling is for the system. This system-level goal must be divided down to the lowest designable or procurable item so that its reliability potential can be considered as a key figure of merit. Reliability allocation provides the method for apportioning the system requirements to each item within the architecture of the system. Allocations are developed using the same functional block diagram that was used to develop the system reliability block diagram and mathematical model. The process starts with the overall system MTBF or failure rate, which then is allocated down to the lower functional levels of the system. The allocation normally is extended down in the hierarchy of the system to the lowest designable item. Allocation is accomplished using past experience of similar systems or best engineering judgments. Figure 5-14 shows how reliability allocations were developed for a reliability block diagram. In this example, the system MTBF requirement is 5000 hours. It has been apportioned to create a design or procurement goal for each lower-level item in the system breakdown. Note that since failure rates are additive, it is easy to see how the total system failure rate was apportioned. During the design of the system, the use of allocated failure rates provides a goal for the reliability of each assembly. As the design develops and the predicted failure rate for each assembly is generated, reliability engineers can compare the predicted failure rates with the allocated failure rates to identify assemblies that will cause the overall system to exceed its required MTBF. Using the comparison as a guide, the reliability engineer can recommend design changes, such as parts with higher reliability, different technologies, or alternate processes, that will achieve the desired level of reliability.

Reliability Predictions

The method for conducting reliability predictions varies with the type of system under consideration and the status of the design. For the purposes of this discussion, two typical methods will be presented to illustrate how reliability predictions are developed. The first method, called *parts-count reliability prediction,* is used in the early stages of design to develop predictions based on the generic failure rates of the anticipated quantity and types of parts to be used in the system. Second is the *part-stress analysis prediction,* which uses the actual design configuration to develop detailed predictions. The parts-count method is valuable in developing initial predictions for comparison with reliability allocations and for providing information to designers and management on the capability of a proposed

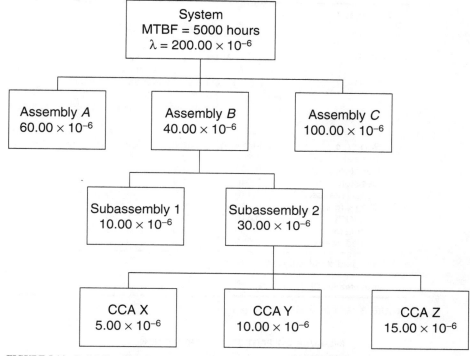

FIGURE 5.14   Reliability allocation.

design to meet basic reliability requirements. Using the parts-stress analysis method as the design matures provides detailed information on reliability and identification of specific areas where reliability can be improved. The validity of both methods depends on the quality of data used in the predictions and the accuracy of assumptions. The prediction of failure rates for items of a system, using either method, begins at the piece-part level and is summed to the system level.

Parts-Count Reliability Prediction.   The parts-count method of predicting system reliability uses a rather simple formula. The formula, illustrated in Figure 5-15, shows that by using generic failure rates from reputable sources, such as actual in-service data or design testing, and anticipated types and quantities of parts, a baseline system failure rate can be derived. The quality factor indicates the level of testing and quality control the part manufacturer used when parts were made. A comparison of the reliability prediction allocated reliability baseline provides visibility of the potential of the evolving design to meet the project reliability goal.

Part-Stress Analysis Prediction.   A reliability part-stress analysis prediction contains a series of complex formulas that are tailored for type of part, and each formula considers the electrical stress, heat, environment, and frequency of use of each part type. This process would be completed for each part for which data did not exist or could not be obtained from the manufacturer. The resulting failure rates then are input to formulas that consider the application of the part and whether it operates in a single, dual, parallel, series, and/or redundant mode. This method identifies individual parts and assemblies that drive the system reliability up or down. Figure 5-16 shows the top-level output of a part-stress analysis reliability prediction. Through use of the part-stress analysis prediction method,

Formula:

$$\lambda_{Equip} = \sum_{i=1}^{i=n} N_i(\lambda_G \pi_Q)$$

Where:

λ_{Equip} = Total equipment failure rate (failures/10^6 hour)
λ_G = Generic failure rate for the ith generic part (failures/10^6 hour)
π_Q = Quality factor for the ith generic part
N_i = Quality of the ith generic part
n = Number of different generic part categories

Part type	Failure rate	Quality	Adjusted	Quantity	Total
Transistor (npn)	0.860	2.0	1.720	8	13.760
Resistors (comp)	0.038	1.5	0.057	2	0.114
Resistors (vari)	0.340	1.5	0.510	6	3.060
Capacitors (cer)	0.170	1.5	0.255	10	2.550
Diodes (gen)	0.140	2.0	0.280	4	1.120
IC (MOS)	0.410	3.0	1.230	12	14.760
Connector	0.150	2.5	0.375	1	0.375
Printed board	0.010	3.0	0.030	1	0.030

Total predicted failure rate ($\times 10^{-6}$)	35.769
Predicted MTBF ($1/35.769)^{10-6}$)	27,957.16

FIGURE 5-15 Parts-count reliability prediction.

MIL-HDBK-217 PART STRESS PREDICTION
Project name : **ABCSYS**
Block 1 : **ABC Computer system**
Description : **Model ACB/XT-8086-based microcomputer**
Analyst : **A.Designer**
Environment : **GB**
Am. temperature : **30**

Part number Description Circuit reference	Blk/Comp F/R Quantity Total F/R	Percent contribution
10 Power supply 110/240 VAC Supply 5V/12 VDC Output	2.33082 $\times$ 1 = 2.33082	6.46
11 CPU Board 8086 Processor + on-board logic	1.61359 $\times$ 1 = 1.61359	4.47
12 Display/memory unit Display processor + RAM/BOM Board	32.12504 $\times$ 1 = 32.12504	89.08

Block F/R = 36.06945 fpm hour
Block MTBF = 27724 hour

FIGURE 5-16 Part-stress reliability prediction.

reliability engineers produce detailed failure rates for all parts, assemblies, and the total system. It must be pointed out that these predictions are just that—statistical predictions. Actual failure rates are not developed until the system is built and then used a sufficient length of time to produce field or test data that are based of actual system usage. Part-stress analysis predictions allow design engineers to ensure that an item meets its allocated goal, which, when added to the predictions of all other items, can be summed to the system reliability goal.

Sneak Circuit Analysis

Design of electronic systems is a very complex task, and there is always the possibility of design engineers overlooking hidden faults in the extensive circuitry of state-of-the-art items. Hidden faults in design can cause unwanted functions or limit performance. The purpose of a sneak circuit analysis is to identify these hidden faults. A sneak circuit analysis is accomplished through a time-consuming circuit-by-circuit analysis of the electronic design. A complete set of engineering drawings and schematics is required. Therefore, the analysis can be done only in the latter stages of design when sufficient detailed documentation exists. The results of the analysis identify undesired circuits, design concerns, or errors in the documentation. A sneak circuit analysis can be very expensive to perform because of the personnel hours required, and given the late stages of design when it is performed, the recommended design changes can be very expensive to implement. Given these constraints, the analysis normally is limited to critical components and circuits.

Tolerance Analysis

Changes in operating temperatures or the effect of different levels over time often will change the electrical operating characteristics of parts and circuits. The purpose of a tolerance analysis is to identify potential occurrences of this condition that will cause the system to exceed its operating specification. This analysis is complex and most often is done using a computer. It is also expensive because of the skill required and the time necessary to develop a computer simulation to do the analysis. Thus it is limited normally to critical items or items that are historically most susceptible to temperature changes, such as circuits using high power levels and power supplies. A tolerance analysis also can address interchangeability of parts from different manufacturers that supposedly perform the same function. The results of the tolerance analysis provide design-change recommendations that should increase system reliability.

Parts-Control Program

The purpose of the parts-control program is to standardize, as much as possible, the parts used in designing a system. This process is to ensure that fully qualified parts that meet standards for reliability and quality are used. The basis for the parts-control program is a preferred parts selection list (PPSL). The PPSL lists all parts that are approved for use on a specific project. This document is extremely valuable on large programs where there may be many subcontractors providing segments of the end item of a system. A customer may provide the contractor an initial PPSL that is amended, with approval, by the contractor as the design evolves. Nonstandard parts must be qualified by the contractor before they can be added to the PPSL. This list is used by design engineers as they select parts for use in the system. There are established data repositories that are used by both governments and industry to accumulate historical information about parts that can be used to assist contractors in selecting parts to reduce the need for nonstandard parts for use in a design. This data is accessible through various procedures. The use of a PPSL on a project fosters selection of parts having established reliability statistics. This makes the results of reliability predictions potentially more accurate and therefore is a valuable step in improving the reliability of a system. This concept is discussed later in Chapters 6 and 11, where the benefits of standardization provide a significant possibility for system design and support.

Reliability-Critical Items Analysis

The purpose of the reliability-critical items analysis is to provide special emphasis on items (parts or assemblies) that are critical to the system achieving its overall reliability goals. These items are the ones that have a high predicted failure rate or that have the most significant potential of causing mission failures. Additionally, these items are of special interest to supportability engineering because they tend to increase support requirements and cost of ownership. Other criteria for being classified as reliability-critical include potential safety hazards, items requiring special handling or transportation precautions, items that are difficult to build, and items with a poor performance history. By increasing the reliability of these critical items, reliability engineers will have the most impact on the total program. Reliability of the system may be increased by adding redundant items, as discussed earlier in this chapter. It also may be increased by use of more reliable parts in an assembly.

Effects of Functional Testing, Storage, Handling, Packaging, Transportation, and Maintenance

Systems must be durable enough to withstand the physical movement and repeated handling and testing to which they will be subjected after they are delivered. The focus of this analysis is to determine the long-term effects that repeated functional testing, storage, handling, packaging, transportation, and maintenance will have on the system being designed. Specific areas of interest include identification of materials or components that deteriorate with age or when subjected to severe environmental conditions, requirements for testing items that are placed in long-term storage, and special procedures that may be required for maintenance or restoration. The results of this task are used in the development of design tradeoffs, field-testing requirements, and plans for packaging, handling, storage, and transportation. Supportability engineers participate extensively in performance of this analysis because it has a direct relationship to support of the system and cost of ownership. Detailed knowledge of the support infrastructure and operational requirements is required to perform this analysis. The results of system requirements identification, as discussed in Chapter 3, are the foundation for this analysis.

Reliability Development and Production Testing

Reliability testing starts with testing parts procured for the manufacturing process and continues through the entire system production. There are four types of testing, listed in Figure 5-17, that address methods for testing parts, assemblies, and/or systems to verify reliability or identify areas

- Environmental stress screening
- Reliability growth test
- Reliability qualification test
- Production reliability acceptance test

FIGURE 5-17   Reliability assessment and test methods.

where improvements in design, materials, or procedures will provide increased system reliability. The use of these tests validates reliability of the final system design and manufacturing process. These tests are applied routinely to all major or critical systems. Supportability engineering may participate in the testing; however, the real importance of these tests in terms of supportability is to ensure that the system will have the necessary reliability potential to minimize support requirements and cost of ownership.

Environmental Stress Screening

The purpose of environmental stress screening (ESS) is to identify and eliminate manufacturing defects from a system. The first testing that can be done to improve the reliability of a system is testing of parts that will be used in the manufacturing process. Procurement of parts that have been screened to a specified set of performance characteristics is the most common method of parts screening. In some cases, the manufacturer may conduct further screening as part of incoming inspection as parts are received from vendors. The next step of ESS is to test assemblies as they are assembled in the manufacturing process. Then the next higher assemblies are tested when assemblies are combined to form systems or subsystems. This process is designed to eliminate any manufacturing defects as early as possible in order to reduce rework or faulty systems to be delivered to the customer. This testing is nondestructive and will identify weak items and workmanship defects and should eliminate early field failures. ESS can be expensive owing to the time required to perform the testing, but it is well worth the expense because it identifies potential reliability problems early, reduces system failures during acceptance testing, and reduces failures during use of the system.

Reliability Growth Test

The purpose of the reliability growth test (RGT) is unique in that this type of testing is conducted to cause the system to fail. The failure then is analyzed to determine if the failure was due to a design defect or error. Using the test-analyze-fix process, RGT is performed on the system or major subsystems under the actual, simulated, or even accelerated environment the system will experience when fielded to identify design deficiencies and defects. The iterative RGT process provides for early incorporation of corrective measures that will provide reliability growth. By subjecting items to test conditions that simulate the anticipated field conditions they will be used under, contractors can identify weak parts, parts that will not perform in the predicted field environment, or design deficiencies that occur when the system is subjected to extreme environmental conditions. This testing, also formerly known as *shake and bake,* applies stress to items in order to identify and eliminate reliability-risk parts. The term *shake and bake* refers to subjecting the parts to vibration and temperature variations while being exercised in realistic situations. Similar tests are performed on assemblies and, finally, the total system. The intended results of this process are to increase operational effectiveness of the system when it is actually used and to reduce the failures that require higher maintenance and logistics support costs.

Reliability Qualification Test Program

As the system design matures and nears time for starting production, reliability qualification test (RQT) is used to verify that the system design meets the performance and supportability goals, thresholds, and constraints established at the start of the project. The preproduction testing is done on one or more samples of the system that are representative of the approved production configuration. RQT differs from RGT in two ways: (1) it is intended to prove the reliability of the system design, not make it fail, and (2) it is used to validate achievement of project requirements. In some cases, RQT is performed by an independent testing agency, not the manufacturer. The results of RQT can be used in making the final decision to authorize the start of full production of the system.

Production Reliability Acceptance Test Program (PRAT)

After production has started, PRAT is used by the manufacturer to verify that systems being produced by the normal manufacturing process continue to meet the reliability performance requirements established by RQT. PRAT is random-lot sample testing. This type of testing is very similar to RQT, except that it is done on full production items. As with RQT, PRAT may be done by an independent testing agency. The quantities of items tested and the frequency of testing depend on the need to verify consistency of the manufacturing process. PRAT can be very expensive to conduct owing to the comprehensive test facilities and time required. The results of this testing may be used for identification and incorporation in design changes if needed to meet reliability requirements.

Failure Reporting, Analysis, and Corrective-Action System

Collection and analysis of reliability data from manufacturing testing and in-service use of a system is invaluable for reliability engineering and supportability engineering. A failure reporting, analysis, and corrective-action system (FRACAS) is a formal method for reporting failures during engineering model testing and then when a system is in operation. FRACAS is a closed-loop reporting system, which means that each reported failure must be followed by a report of the failure analysis and the corrective action required. Figure 5-18 is a typical FRACAS flow diagram. The FRACAS process is initiated by the manufacture or procuring organization and continues to operate through the operational life of the system. The FRACAS program focuses on long-term improvement requirements for the system or tracking of failure trends when the system is in operation. The failure data gathered by a FRACAS process form the basis for reliability engineering as inputs to predictions of failure rates for the next system. Supportability engineering uses FRACAS data to refine projections of system availability, supportability, and cost of ownership.

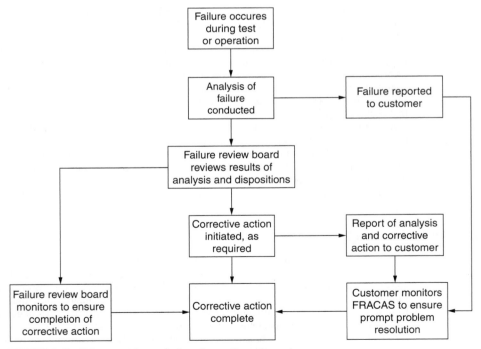

FIGURE 5-18   Failure reporting, analysis, and corrective-action system.

The purpose of reliability engineering is to aid in the design of a system to ensure that it works as long as possible without failing. Additionally, it predicts statistically how and when the system will fail. Reliability engineers work hand in hand with design engineers throughout the design of a system. The results of the reliability program are used extensively by supportability engineering in identifying recommendations for improvement of system supportability and in determining the logistic support resources that will be required to sustain the system once it is in operation. As will be illustrated in subsequent chapters, the accuracy of reliability predictions and the FMECA drive requirements for quantities of support resources, which, in turn, determine the projected life-cycle costs of the system.

MAINTAINABILITY ENGINEERING

How do you design a system that is easy, quick, and cost-effective to repair when it fails? Reliability engineering, as discussed earlier, participates in the design process to produce a system that is as failure-free as possible. Maintainability engineering also participates in the design process, but its effort is focused on making system as easy and inexpensive as possible to test, service, and repair when it does fail. The goals of maintainability engineering are to provide input into the design process in the form of design criteria that result in a system design in which it is easy to identify faults, requires minimum personnel and other logistics support resources to perform maintenance, and has the lowest life-cycle cost possible. The concepts used by maintainability engineering to accomplish these goals, such as modeling, allocations, predictions, and testing, are very similar to those used by reliability engineering. Many of the results of reliability activities are used by maintainability engineers, which will become evident later in this chapter.

The initial task of maintainability engineering is to develop predictions of how long it will take to repair the system when it fails. Using these predictions, the design can be analyzed to identify possible changes that would reduce the time required to perform maintenance. As the design matures, the maintainability aspects of the system can be determined through actual testing and demonstration of maintenance actions. The combination of system reliability, how often it will fail, and system maintainability, how long it takes to repair a failure, has a direct correlation with the amount of time that the system will be capable of performing its mission. Fewer failures require less time for maintenance, so the more time the system is operational. Given this fact, it is not uncommon to see these disciplines referred to together as *reliability and maintainability* (R&M).

Definitions

An understanding of the following terms used in this chapter will aid in understanding maintainability engineering:

Maintainability   The probability that a failed item can be repaired in a specified amount of time using a specified set of resources. Note that this is a statistical prediction, which means that, like reliability, maintainability can be greatly influenced by variables such as availability of resources and environmental conditions where maintenance is performed. Also notice that this definition assumes that a specified set of resources will be available to support the repair process. The actual resources required for repair will be addressed in subsequent chapters.

Maintenance   The overall physical activity established to restore an inoperable item to an operable condition or that is intended to prevent an impending or future failure of an item. Maintenance typically is divided into corrective or unscheduled maintenance and preventive or scheduled maintenance. Maintenance requirements also may be described by where they will be performed, such as on-equipment and off-equipment.

Maintenance task   Any single action or sequence of actions that restores an inoperable item to an operable condition or prolongs the serviceability of an item.

Mean time to repair (MTTR)   The average time required to perform maintenance over a specified operating period. Initially, the MTTR is developed using predicted times to perform maintenance tasks. When the design is complete, the MTTR can be refined by actually measuring the time to perform tasks. The accuracy of the MTTR then depends on the correctness of reliability predictions and the accuracy of determining exactly what is required to perform maintenance. The MTTR is developed during development. It is assumed that all resources required to perform maintenance are available. The MTTR represents the best-case scenario for performing maintenance. In the real-world, in-service use of the system, the actual amount of time typically is longer.

Maximum time to repair (MAX_{TTR})   The maximum allowable time required for any single on-equipment maintenance task. The MAX_{TTR} creates a limitation on the overall time required for performing on-equipment maintenance.

Mean corrective maintenance time (Mct)   The average time required to perform corrective maintenance based on actual in-service experience.

Mean preventive maintenance time (Mpt)   The average time required to perform preventive maintenance based on actual in-service experience.

Mean time between maintenance (MTBM)   The average time between performance of all maintenance actions. MTBM includes corrective maintenance resulting from inherent failures, corrective maintenance resulting from maintenance-induced errors, false removals, and preventive maintenance.

Mean man-hours per maintenance action (MMH/MA)   The average number of man-hours required to perform a maintenance action. The MMH/MA is used to develop a prediction of the total quantity of labor that will be required to perform maintenance. Multiplying the MMH/MA of a system by the predicted number of failures of that system during a specific amount of time results in the anticipated number of man-hours that will be required to perform maintenance.

Mean man-hours per operating hour (MMH/OH)   The ratio of man-hours required to perform maintenance to 1 hour of system operation. This number is used as a gauge to develop tradeoff comparisons between different ways of doing maintenance. When developing the MMH/OH statistic, it is important to ensure that the maintenance tasks considered in the calculation are appropriate. Normally, this calculation includes all corrective and preventive maintenance tasks done on the system, but off-system maintenance may be included in some special cases.

Note: This definition employs the use measurement base of hours; however, it could be miles, rounds, landings, or any other measurement base.

Fault detection   The act of identifying that a fault has occurred in a system. This may be done by inspection, testing, or other means.

Fault isolation   The act of identifying a failure to the level that will enable corrective maintenance to begin. In other words, doing some kind of testing or evaluation to get down to the level where something can be replaced or repaired in order to fix the problem.

Concept of Maintainability

Maintainability focuses of the design characteristics of a system that will enable it to be maintained as efficiently and cost-effectively as possible so that it is available to perform assigned missions. Maintainability, just like reliability, generates and uses statistics to project what might happen in the future. The primary maintainability statistic at the start of a development or procurement project is mean time to repair (MTTR). This statistic is a critical ingredient to estimating the amount of time that a system will not be available owing to performance of maintenance. The MTTR is a weighted-average amount of time required to perform any maintenance task. The definition for maintainability provided earlier includes two significant phrases: (1) a specified amount of time and (2) a specified set of resources. The specified amount of time is the MTTR. The specified set of resources is a little more complex. This relates to differing levels of maintenance capability, where varying quantities of resources may be available to support the system.

Typically, on-equipment maintenance is limited to actions that can be done quickly in order to return a failed system to an operable condition. Therefore, it should be anticipated that a limited range of resources would be available for on-equipment maintenance. Items will be removed from the system and sent to dedicated maintenance locations for repair. A dedicated maintenance location would be expected to have a far greater range of resources available. Thus maintainability performs its activities in anticipation of these facts—that various maintenance capabilities, such as on-equipment and off-equipment items, will have different types and quantities of resources available to maintain a system. Identification of the existing and planned future support infrastructure was discussed in Chapter 3. The system requirements study provides a mechanism for recording this information. This information is the starting point for maintainability to define what the specified set of resources may be simply by stating that maximum use will be made of existing resources.

Maintainability Statistics

The MTTR for a system is the key figure of merit for establishing a benchmark for maintainability. As stated previously, it is the weighted-average time required to perform any maintenance action at a given level of maintenance. Calculation of the MTTR for a system combines the inherent failure rate of an item with the time required to perform a maintenance task to restore it to an operable condition. Before a detailed discussion of the derivation of MTTR, an understanding of maintenance is required.

A corrective maintenance action that restores a system to an operable condition typically consists of a fairly standard series of events known as the *corrective maintenance cycle*. Figure 5-19 illustrates this series of events. Notice that the cycle begins when the failure is detected, not when it occurs. The preparation for maintenance may consist of doing some type of fault-verification inspection or test to determine that corrective maintenance actually is required. This step also may include use of diagnostic

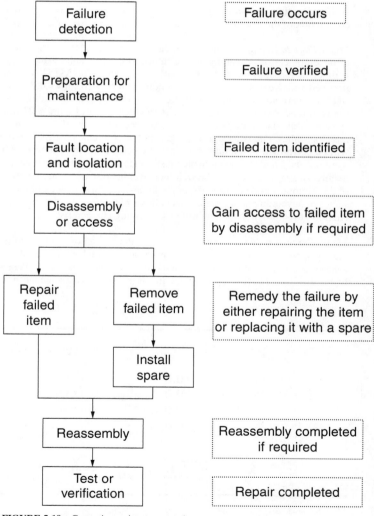

FIGURE 5-19   Corrective maintenance cycle.

$$\text{MTTR} = \frac{\Sigma \,(\lambda_I \times \text{maintenance task time})}{\Sigma \,\lambda_I}$$

FIGURE 5-20   MTTR calculation.

test equipment for some items. Frequently, the failed item is inside, behind, or underneath another item, so some disassembly or removal of another item may be required to gain access to the failed item. Then, to actually fix the problem, either the failed item is removed and replaced with a spare, or the failed item is repaired in place. After the failure has been rectified, the items that were opened or removed to gain access to the failed item must be reassembled. The final step is to perform a test or verification that the maintenance action was successful. The corrective maintenance cycle ends at completion of the test or verification.

Mean Time to Repair.   The mean time to repair (MTTR) is a statistic developed in the early phases of development or procurement to establish goals for maintenance and to identify potential critical maintenance requirements for the evolving system. The MTTR is a design figure of merit that is imposed on the design engineer for an item. Figure 5-20 shows the generic formula used to calculate MTTR. There are two inputs to this calculation, the inherent failure rate of each item and then the time required to repair or remove the item when it fails. Figure 5-21 illustrates how the MTTR for a system is calculated. In this example, the system has three boxes. Since each box fails at a different rate, the failure rate is used as a weighting factor to calculate the frequency that a maintenance task will be performed in relationship to maintenance tasks for other items in the system. If this method were not used and the maintenance task times simply were averaged, the result would be approximately 41.7 minutes. Use of the weighted average provides a more realistic maintenance requirement of 33.5 minutes average time for any maintenance task. Calculation of a system MTTR in the early stages of development or procurement allows identification of critical areas of the design that require attention to improve maintainability by changing characteristics to lower maintenance requirements.

Mean Corrective Maintenance Time.   As the design of the system matures, more detailed information about maintenance requirements becomes available. The actual time to perform corrective maintenance allows a more complete description of the corrective maintenance cycle for each item. Eventually, the MTTR is replaced with the mean corrective maintenance time (Mct). The Mct considers

Box	λ_I	Task time	$\lambda_I \times$ Task time
Box 1	0.006	45 min	0.27
Box 2	0.011	20 min	0.22
Box 3	0.003	60 min	0.18
	0.020		0.67

$$\text{MTTR} = \frac{0.67}{0.020} = 33.5 \text{ minutes}$$

FIGURE 5-21   MTTR example.

other factors such as access time and preparation for maintenance that may have been either estimates or not considered when the MTTR was developed. This leads to a more realistic portrayal of the actual requirement for maintenance when the system is delivered to the user. The various people who actually perform maintenance typically have varying levels of proficiency, and the conditions under which maintenance is performed may vary depending on location, weather, and other factors. In-service maintenance data collection, over time, will report the number of times that a specific corrective maintenance cycle occurred and the actual time required for each iteration of the cycle. These times can be averaged to derive the Mct.

Mean Preventive Maintenance Time.   Preventive maintenance actions are to preserve operability of the system or to address a fault before it progresses to a failure. Preventive maintenance often consists of inspections, condition monitoring, and scheduled removal of selected items subject to wearout or time-dependent degradation. The requirement to perform a preventive maintenance task is determined through a process called *reliability-centered maintenance* (RCM). This process looks for ways to determine that a failure is about to occur so that the preventive maintenance action is triggered to be performed. The frequency of performing a preventive maintenance action is determined by the criticality of the failure the action is intended to prevent. Typically, preventive maintenance actions are performed based on calendar days (such as daily, weekly, or monthly) or based on usage events, such as after a predetermined number of aircraft landings, miles driven, or hours of operation. Preventive maintenance actions vary based on the technology of the system. Maintainability engineering uses the RCM methodology to determine specific preventive maintenance actions for a system and then determines the frequency of performance. After a preventive maintenance action is identified, it is analyzed to estimate the amount of time required for its performance. Mean preventive maintenance time (Mpt) then is calculated using the frequency of performance as a weighting factor just as the failure rate was used as a weighting factor for corrective maintenance.

THE MAINTAINABILITY PROGRAM

The goal of the maintainability program is to improve the operational availability of system while reducing the requirements for personnel and other logistics resources. The methods used by maintainability engineers to achieve this goal are similar in methodology to reliability engineering. The methodology consists of a series of analyses and then testing and assessment, as shown in Figure 5-22.

- Maintainability modeling
- Maintainability allocations
- Maintainability predictions
- Failure modes, effects, and criticality analysis
- Design criteria
- Maintainability/supportability demonstration
- Data collection, analysis, and corrective action system

FIGURE 5-22   Maintainability engineering activities.

Maintainability Modeling

The use of models for determining the maintainability of a system forms the basis for having the ability to predict the ease with which maintenance will be performed on the system. They should be developed early in the concept phase. Through use of an appropriate model, maintainability engineers can develop an initial prediction of system maintainability and subsequently determine the effect of changes on the total system and evaluate the need for redesign to achieve maintainability goals. The selection or tailoring of the appropriate model to be used to determine the maintainability characteristics of a system is the first step in predicting the number of hours that the system will be unavailable for use. The models that are provided in the handbook must be tailored to reflect the appropriate configuration and complexity of the system.

Figure 5-23 illustrates the model used to predict the MTTR of an item. The model for MTTR prediction uses the common tasks required to repair a failed item. Each of the elements of the model likewise must be predicted using models, so a submodel for each subelement must be developed. The complexity of the system being modeled dictates how detailed a model should be. Small items of the system may require only selected portions of the model shown in Figure 5-23, whereas large, complex systems require extensive submodels to determine the input for the overall system model. If a system has several distinct levels, such as a tank, where the system has some assemblies that are primarily mechanical, others that are electronic, and some that are a combination, then the model for the system may be composed of submodels that are tailored for each type of assembly. This process of developing an accurate and appropriate model will determine the validity of the resulting predictions. In each case, the model must be as representative as possible of the task functions required to maintain the system. As with all models, the resulting predictions form a baseline for comparison of recommended changes to determine their effect on the overall system. This modeling is used to develop early maintainability goals, thresholds, and objectives for a project.

Maintainability Allocations

The maintainability design process begins with development of an allocation of the overall system maintainability goals. The maintainability goals of a system normally are found in the procurement specification. These goals may be expressed as MTTR figures for each level of maintenance (e.g., MTTR at organizational level will not exceed 0.25 hour and at intermediate level will not exceed

$$\text{MTTR} = \overline{T}_p + \overline{T}_{FI} + \overline{T}_{FC} + \overline{T}_A + \overline{T}_{CO} + \overline{T}_{ST} = \sum_{M=1}^{m} \overline{T}_M$$

where $\overline{T}_p$ = Average preparation time

$\overline{T}_{FI}$ = Average fault isolation time

$\overline{T}_{FC}$ = $\overline{T}_D + \overline{T}_I + \overline{T}_R$

$\overline{T}_D$ = Average disassembly time

$\overline{T}_I$ = Average interchange time

$\overline{T}_R$ = Average reassembly time

$\overline{T}_A$ = Average alignment time

$\overline{T}_{CO}$ = Average checkout time

$\overline{T}_{ST}$ = Average startup time

$\overline{T}_M$ = Average time of the Mth element of MTTR

FIGURE 5-23 Maintainability model.

0.75 hour). Another method for expressing maintainability goals is as a ratio of system operating hours to maintenance hours [e.g., mean maintenance hours per operating hour (MMH/OH) will not exceed 0.005]. For design planning purposes, these goals represent the maximum mean time that can be allowed for maintenance at a given maintenance level. The contractor begins the allocation process using these goals, which are allocated down for each lower system level. The maintenance concept for the system plays an important part in the allocation process. For example, if the maintenance concept calls for only two levels of maintenance, then the maintainability allocation should consider only those levels. Additionally, the maintenance concept may establish criteria for the maintenance actions that will be accomplished at each level. If the maintenance concept states that organizational maintenance will consist of removal and replacement of modules, then maintainability considerations at the organizational level should be centered on improving the accessibility and ease of removing and replacing modules. The allocation of maintenance time should reflect this maintenance concept.

Figure 5-24 illustrates allocation of an MTTR from system level down two indentures. These allocations must be provided to the design engineers responsible for designing in the maintainability of the system so that they understand the requirements for maintenance that have been levied on their area of responsibility. The allocation process also provides a bookkeeping procedure for establishing a baseline for achieving the system maintainability goals. As the design matures, individual problems in achieving the goals can be evaluated to determine their effect on the overall system. This allows continual evaluation of system maintainability in relation to the goals and allows tradeoff analyses to determine where changes will have the most benefit. The allocation process also aids contractors in achieving system goals when dealing with subcontractors. By allocating the overall system goals down to levels that are supplied by subcontractors, the contractor can manage the total maintainability program rather than having to rely on chance to achieve the contractually stated maintainability goals. The initial allocation process should be based on historical information about the maintainability of similar system, initial predictions, or best engineering judgments. The allocations may change as the design matures, but the goals cannot be changed without concurrence of the customer. As shown in the example provided, the maintainability allocation applied to each item is heavily dependent on the corresponding reliability statistic for that same item and its next higher assembly.

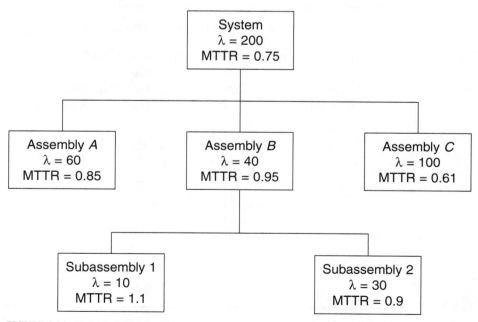

FIGURE 5-24   Maintainability allocation.

Maintainability Predictions

The purpose of a maintainability prediction is to determine, using mathematical calculations, if a system design will meet the established maintainability goals. The prediction process is also used to identify designs that will not meet the goals. During the early phases of a development or procurement project, sufficient information is not available to perform detailed prediction calculations, so predictions may be based on the performance of previous system using design changes to modify the result to closely relate to the new design. In order to perform detailed calculations, information such as the maintenance concept, functional block diagrams, identification of replaceable units, and reliability estimates is necessary. Using this information, maintainability engineers can develop predictions for the system. As the design matures, the quantity and quality of information increases, which allows further refinement of early predictions.

Maintainability Analysis

Maintainability analysis is one of the primary parts of the maintainability program. The purpose of this analysis to identify system maintainability design features that will enable the system to meet maintainability goals, evaluate design alternatives, provide detailed input to the maintenance planning process, and continually evaluate the design to ensure that goals are met. This analysis has a significant overlap with analyses being conducted simultaneously by other disciplines, and close coordination is required to avoid duplication of effort. Maintainability engineers require detailed information on system design and other information such as the maintenance concept and plan, anticipated test and fault-isolation capabilities, maintenance skills, operational information, and reliability predictions to perform a usable analysis.

Development of statistics such as mean maintenance hours per operating hour (Figure 5-25) can be extremely beneficial in reducing requirements for operation and personnel. The maintainability analysis process develops detailed design criteria that are necessary to meet established goals.

Formula: MMH/OH = (MTTR $\times$ C $\times$ F)/MTBF

Where: MTTR = Mean time to repair
   C = Crew size
   F = Operation service ratio
   MTBF = Mean time between failures

Item: Name: Radar set
   Number: 12345-1

Date: MTTR = 0.25
   C = 2
   F = 1.5
   MTBF = 1500

Results: MMH/OH = 0.0005

FIGURE 5-25   Mean maintenance hours per operating hour.

These criteria can consist of requirements for accessibility, tool usage, test points, standardization of tools and procedures, connectors and fasteners, and modularization. These criteria serve as guidelines for engineers as the system is designed. As the design evolves, alternatives are identified and evaluated to determine the approach that provides the best maintenance capabilities. These alternatives must be evaluated as early as possible to allow implementation without unnecessary redesign that would occur later in the program phases. The maintainability analysis is a key element in selecting the most desirable methods for testing the system when a failure occurs. The adequacy and efficiency of the system testability has a large impact on the maintainability of a system. The easier it is to find a failure, the quicker it can be fixed. The clock starts for the corrective maintenance cycle when the failure is detected and does not stop until the system is repaired. Excessive time requirements for testing can only increase the overall system down time and degrade the system maintainability.

Inputs to the maintainability analysis come from reliability analyses and predictions, human factors engineering information of recommended skill levels and quantities of maintenance personnel, system safety analyses, and maintenance planning. Outputs from the maintainability analysis are used as inputs to other maintainability activities and by supportability engineering for quantifying the support infrastructure required for the system.

Maintainability Design Criteria

The maintainability engineering activity on a development program establishes the ability to quantify the criteria for system design to achieve established maintainability goals. This takes the results of previous maintainability tasks and translates requirements into guidelines that can be used by design engineers during system design. Figure 5-26 provides examples of maintainability design criteria. These criteria commonly are included in the product specification as critical requirements for

- Assemblies and repair parts having the same part numbers will be functionally and physically identical.
- Access to maintenance-significant items will be provided through entries that do not require removal of other components.
- Scheduled maintenance, alignments, and calibration requirements will be avoided.
- Special tools or test equipment will not be required to perform maintenance at organizational- or intermediate-maintenance levels.
- Captive, quick-release fasteners will be used to secure maintenance access panels or covers.
- All screws and bolts will be of standard dimensions to reduce tool requirements.
- Results of maintainability predictions.
- Unscheduled and scheduled maintenance requirements.
- Tool and test equipment requirements.
- Skill-level requirements for each maintenance level.
- Fault identification requirements.
- Types of maintenance tasks required at each level.
- Test requirements for fault isolation.

FIGURE 5-26 Maintainability design criteria.

design of the product. An interpreted or expanded version of the design criteria is developed through the systems engineering process as detailed "how to" instructions as to physical attributes to be included in the final design to meet the maintainability requirements of the program.

Maintainability Demonstration

Proof of the effectiveness of the maintainability program is determined through the accomplishment of maintainability evaluation and testing. This evaluation and testing are necessary to verify that the maintainability design criteria have been incorporated into the system design and will result in achievement of the stated maintainability goals. Historically, this evaluation was termed a *maintainability demonstration;* however, it is rapidly being expanded to a *complete supportability demonstration* that evaluates both the system design and the resource infrastructure that is to be in place to support the system's operation. The purpose of this demonstration is to physically show that the system is capable of being maintained. The process for conducting the demonstration is relatively simple. It is conducted using the actual system late in the development phase, normally just prior to the start of full production. Also required are the technical manuals, tools, and other support system necessary to accomplish maintenance. It should be very evident that this demonstration affects not only maintainability but also virtually all supportability engineering.

The object of the demonstration is to take an operational system, induce failures into the system, and use only the technical manuals and support system that will be available to maintenance personnel when the system is fielded to find and fix the failure. It is best if actual service technicians perform the maintenance rather than engineering or manufacturing personnel to get a true picture of the maintenance of the system. In other words, this demonstration is a test of how well the developer has designed the system for maintenance and how usable the logistics support package is in actually performing maintenance. The maintainability demonstration is a significant project milestone that occurs close to the end of development. Failure to pass the demonstration may result in system redesign, technical manual changes, and delays starting full production of the system until all the problems identified during the demonstration are remedied.

Data Collection, Analysis, and Corrective-Action System

Every development program generates more data than could ever be digested and used to its fullest extent without specific guidance as to what should be collected and what to do with it after it is collected. Relevant maintainability data is derived from maintainability analyses, engineering tests, demonstration tests, and user tests. This data is compiled and analyzed to develop trends and identify critical areas for further investigation to resolve design problems or enhance overall performance. It is important to understand that this maintainability data is also the same base information that is used by supportability engineering. The format and method for compiling these data should produce sufficient information to identify maintenance actions, time required, number of personnel required, adequacy of the support and test system, methods for fault detection and isolation, and any special or unusual occurrences.

The maintainability program provides a unique input to the design process and development of the final support infrastructure for the system. Whereas the reliability program seeks to make the system as failure-free as possible, the maintainability program takes the approach of assuming that the system will fail and addressing how to best design the system so that it can be fixed when it does fail. The programs must be coordinated to receive maximum benefit from both viewpoints. Supportability engineering is a key participant in ensuring that this coordination occurs.

TESTABILITY ENGINEERING

An integral part of the overall design effort of electronic systems is testability engineering. The reason for this importance is that testability engineering addresses the requirements for testing that must be considered in the development and design of an electronic system or systems. This includes the extent to which a system or system design supports fault detection and fault isolation in a confident, timely, and cost-effective manner. Historically, testability began as a subset of the maintainability engineering process. However, in recent years, electronic designs have become so complex, especially with multiple layers of redundancy, that testability has been established as a separate and distinct engineering activity. Testability engineering is intimately involved with both reliability and maintainability.

As stated previously, reliability engineering focuses on making the design as failure-free as possible, and maintainability focuses on making the design as easy and cost-effective to repair as possible when it does fail. Testability is actually a bridge between reliability and maintainability. Before the failure can be fixed, it must be identified. And before testing can be planned, the possible failures that may occur in a system must be predicted. Therefore, testability is the link between reliability, where possible failures are identified, and maintainability, where the failures are fixed. The degree of testability exhibited by a system design is in direct correlation with achievement of maintainability goals. This is accomplished through incorporation of adequate testability, including built-in test, into the design. The easier and quicker faults are located and identified, the more rapid maintenance can be accomplished.

However, testability must be applied with reason to any design. Excessive built-in test capability in a system design can cause a significant increase in system costs which may not be justified. Therefore, testability requires early and systematic management attention to testability requirements, design and measurement in order to achieve a balance between meeting reliability, maintainability, testability, performance, and cost.

Testability   A design characteristic that allows the status (operable, inoperable, or degraded) of an item to be determined and the isolation of faults within the items to be performed in a timely manner.

Built-in test (BIT)   An integral capability of the mission system or system that provides an automated test capability to detect, diagnose, or isolate failures. Normally, BIT is performed by software that runs on operational hardware.

Built-in test equipment (BITE)   Hardware and software that are identifiable as performing the built-in test function but have no operational functionality.

False alarm   A fault indicated by BIT or other monitoring circuitry where no fault exists.

Fault-isolation time   The elapsed time between the detection and isolation of a fault.

Cannot duplicate (CND)   A fault indicated by BIT or other monitoring circuitry that cannot be confirmed.

Fault detection rate   The ratio of failures detected by BIT or other testing procedures to the failure population.

Fault detection time   The time that elapses between the occurrence of a fault and the detection of the fault by the test process.

Fault resolution   The degree to which a test program or procedure can isolate a fault within an item; generally expressed as the percent of the cases for which the isolation procedure results in a given ambiguity group size.

Off-line testing   The testing of an item when the item has been removed from its normal operating environment.

Retest okay   A unit under test that malfunctions in a specific manner during operational testing but performs that specific function satisfactorily at a higher-level maintenance facility.

Test effectiveness   Measures that include consideration of hardware design, BIT design, test-system design, and test-program set design. Test effectiveness measures include, but are not limited to, fault coverage, fault resolution, fault detection time, fault isolation time, and false alarm rate.

Testability Goals

The goals of testability engineering are to influence the system design to make the final product as testable as possible. When applied appropriately to a developmental program, the testability program has the potential of significantly increasing the user's confidence and satisfaction in the system. These basic testability engineering goals are shown in Figure 5-27. By increasing the ability to test

- Facilitate the development of high-quality tests.
- Facilitate manufacturing test.
- Provide a performance-monitoring capability.
- Facilitate development of fault-isolation procedures for technical manuals.
- Improve the quality and reduce the cost of maintenance testing and repair at all levels of maintenance.

FIGURE 5-27   Testability engineering goals.

the system, whether to verify that it is operational or to identify a failure, the user saves time and resources in operation and maintenance of the system.

Test Effectiveness Measures

During the detailed design process, testability engineers measure the effectiveness of the test capability being included in the overall system design. The statistical information provided by these effectiveness measures allows the system design engineers to gauge how well testability actually is being included in the design. The most common test effectiveness measures are fault coverage, fault resolution, fault detection time, fault isolation time, and system-level test effectiveness. Formulas for determining fault coverage, fault resolution, and system-level test effectiveness are provided in Figures 5-28 through 5-30. The results of each of these calculations provide an indication of how well testability requirements for the system are being met and should be used to highlight areas of the design that are suspect and require further analysis to identify methods for increasing inherent testability. Fault detection time and fault isolation time are significant inputs to the maintainability program for determining the length of time required to perform maintenance actions.

$$FD = \frac{\lambda_\delta}{\lambda}$$

Where:

$$\delta = \sum_{i=1}^{K} \lambda_i$$

$\delta = K\lambda_i$

λ_i = Failure rate of ith item

K = Number of detected failures

FD = Fault-detection rate

FIGURE 5-28   Fault detection coverage formula.

$$FR = \frac{100}{k} \left(\sum_{i=1}^{N} m_i \right)$$

where K = Number of detected faults
        N = Number of unique failure responses
        i = Failure response index
        M_i = Number of modules in failure response
        FR = Fault resolution rate

FIGURE 5-29 Fault resolution formula.

$$\frac{\text{System test}}{\text{Effectiveness}} = \frac{\sum \lambda_i FD_i}{\sum \lambda_i}$$

where

     λ_i = Failure rate of ith item
     FD_i = Failure detection prediction

FIGURE 5-30 System test effectiveness formula.

TESTABILITY ENGINEERING PROGRAM

The testability program for a system design effort must be integrated with other design and analysis tasks to achieve the goals of the design program. The goals of a testability program are shown at Figure 5-31. The first goal of the testability program is to support the maintainability program in the areas of meeting performance monitoring of the system and designing into the system the capability of testing for corrective maintenance actions at all levels of maintenance. Second, the testability program must provide a basis for the logistics support planning effort with regard to selection and use of support and test systems. The third goal of the testability program is support of and integration with the design engineering effort to meet testing requirements, including the hierarchical development of testability designs from the piece part to the total system.

- Integration with system and design engineering to meet testability requirements
- Support of and integration with maintainability design
- Support of logistic support requirements planning

FIGURE 5-31 Testability engineering goals.

Testability Requirements

The first step in meeting the goals of a testability program are to determine the discrete testability requirements for the system being designed. The desired result is to determine the best alternative testability requirements for the new system when considering the effectiveness of the testing, the depth of testing required to support maintenance, and the cost tradeoffs of manual versus automatic testing. The development of test alternatives should consider the use of new technologies that enhance the testability of the system being designed. Additionally, maximum use should be planned for existing test capabilities in order to reduce the need for development of a new test system that would have an impact on the life-cycle cost of the system. The testability requirements must support the maintenance concept with the minimum test capability necessary to accomplish maintenance.

For example, suppose that the maintenance concept for a system being designed called for failures at the organizational level to be repaired by removal and replacement of major subsystems. Then the testability requirements for organizational level should be developed to support identification of subsystems that fail and testing of subsystems for operational condition, nothing more. Although in most cases BIT could be included to isolate the faults to a lower level than the major subsystem, to do so would incur unnecessary costs because such detailed BIT is not required to support maintenance. On the other hand, an alternative maintenance concept could be developed that includes repair of major subsystems by removal and replacement of failed assemblies rather than removal of the complete subsystem. This creates a requirement for a tradeoff analysis that would consider all the parameters of both concepts to determine the most desirable. The results of the tradeoff analysis must be supported by both quantitative and qualitative information germane to the maintenance concepts. The key issue is achieving a balance among the performance, supportability, and cost of the design. Overdesign incurs unnecessary costs that should be avoided, and underdesign creates a system that will not meet specification requirements.

There are three basic design areas that testability is concerned with, especially during the preliminary design stages. These areas are test-system compatibility, BIT design, and the physical structure of the design. As will be discussed in later chapters, customer organizations normally maintain a standard inventory of test systems used to support maintenance. Planning for use of the support system selected from this inventory for testing eliminates the need for special test systems that have limited application. By using this standard inventory of test systems, the long-term cost of ownership for the customer can be reduced significantly. Any time that a special test system is developed for a single application, the cost associated with development and support of the test system causes the total cost for the prime system to escalate. By planning early in the design process for the system to be testable with a standard test system, excessive test system costs can be avoided. In addition to reducing costs for the test system, planning for compatibility with the test system is a major task early in the design process. By designing the system so that it can be tested easily using a standard test system, with test points located in strategic locations, redesign for after-the-fact test compatibility also can be avoided.

BIT planning is a consideration that must be included early in the design process. Sometimes it may be very difficult to perform BIT analyses so early owing to a lack of definitive design information; however, in order to gain the maximum benefit from BIT, it must be included upfront. The basis for BIT planning and analyses should be a determination of the probable faults that are predicted to occur when the system becomes operational. The source for this information is the FMECA performed by reliability engineering. Since the FMECA identifies all the predicted failure modes of the system, it is the logical starting point to plan for identification of the faults at some level using BIT.

The last preliminary design area that testability addresses is the physical structure of the system. Development of a testable system requires that the physical layout of the system be such that it allows maximum access for testing. Functional partitioning is the single most effective method for enhancing testability in a design. Each function should be partitioned in a method that isolates the function for testability purposes. This increases the ability to test using either BIT, test points, automatic testing equipment (ATE), or manual testing.

Test Design Tradeoffs

The overall test design of a system includes a combination of manual testing, BIT, and off-line automatic testing. Early in the preliminary design of the system, two basic testability tradeoffs must be accomplished, the results of which guide all testability design-related activities thereafter. The first testability tradeoff is to determine what testing will be done manually and what tests will be accomplished using ATE. Choosing the appropriate test method, manual or automatic, requires information from the FMECA, maintainability analyses, support and test system analyses, and the maintenance concept. The second tradeoff is to determine which automatic testing methods will be used, BIT or off-line automatic testing, to identify and isolate faults. If the product specification states that all organizational-level testing will be accomplished using BIT and that the organizational level will be limited to removal and replacement of failed assemblies, then BIT must be able to identify faults at the assembly level. Having a BIT capability of locating faults within assemblies is an excessive amount of BIT because, based on the maintenance concept, it would never be used. Off-line automatic testing using ATE at different locations from where the system is used operationally allows maintenance to be performed to a more extensive level than at the organizational, or first, level. A combination of manual and automatic testing may be required at off-line maintenance facilities to repair an item completely.

The final combination of manual, BIT, and off-line testing is determined by which amount of each provides the best balance among performance, support, and cost. While it is technically possible to design a system that is completely testable using BIT, it is impractical owing to cost, space, and impact on the functional requirements of the system. Conversely, while any fault could, in theory, be identified using only manual testing, limiting test capability to manual processes is not logical owing to the time, training, and documentation required to support maintenance. This is why a combination of all three provides the best of all possible alternatives.

Inherent Testability Assessment

Assessing the inherent testability of a system design provides feedback to testability engineers for identification of areas of the design that require further analysis or design changes so that the final system design can meet testability goals. There are many methods for qualitatively assessing the testability of a design. Figure 5-32 contains a checklist that, when tailored for the specific system being designed, provides a logical and comprehensive evaluation tool for assessing the inherent testability of a design. Whether this checklist or other methods are used to assess inherent design testability, the important point is that some logical and objective method must be used to guide the effort of design engineers early in the design process in order to result in a testable system design.

Testability Cost and Benefit Data

The testability program and incorporation of testing capabilities into a system design are not without a significant cost. Therefore, there must be a balance between the cost incurred for test capability and the benefit received from the level of testing incorporated in the design. Too little spent on designing in test capability may result in years of wasted resources in performing testing that could have been incorporated in the original design. Conversely, there is a point of diminishing returns where no value is gained from incorporating additional test capability. The key is to be able to weigh the costs and benefits and achieve a balance between both so that the resulting testability of a system provides the best testing for the minimum long-term expense. Figure 5-33 provides a listing of the points that must be considered when determining how much test capability is the right balance between cost and benefit.

Test and Evaluation

The formal test and evaluation of the testability of a system design occurs as an integral part of the maintainability demonstration described previously in this chapter. The purpose of the maintainability demonstration is to demonstrate to the customer that the final qualified system design can be maintained with the logistics support package—tools, test system, training, and technical manuals—that the contractor has developed. The first step in conducting the maintainability demonstration is to induce controlled faults into a known good system and use BIT or ATE to locate the fault. This is

Mechanical design

1. Is enough spacing provided between components to allow for clips and test probes?
2. Is a standard grid used on boards to facilitate component identification?
3. Are all components oriented in the same direction?
4. Are standard connector pin positions used for power, ground, clock, etc., signals?
5. Are power and ground included in the I/O connector?
6. Are connector pins arranged such that shorting of physically adjacent pins will cause minimum damage?
7. Is each hardware component clearly labelled?

Partitioning

1. Is each function to be tested placed wholly on one board?
2. If more than one function is placed on a board, can each be tested independently?
3. Within a function, can complex digital and analog circuitry be tested independently?
4. If required, are pull-up resistors located on the same board as the driving component?
5. Are analog circuits partitioned by frequency to ease tester compatibility?

Test control

1. Can circuitry be quickly and easily driven to a known initial state?
2. Is it possible to disable on-board oscillators and drive all logic using a tester clock?
3. Is circuitry provided to bypass any one-shot circuitry?
4. Are active components used to allow the tester to control necessary internal nodes using available input pins?
5. Are unused connector pins used to provide additional internal node data to the tester?
6. Are buffers or divider circuits employed to protect those test points which may be damaged by an inadvertent short circuit?

Parts selection

1. Is the number of different part types used in the design the minimum possible?
2. Have parts been selected which are well characterized in terms of failure modes?
3. Is a single logic family being used? If not, is a common signal level used for interconnections?

Analog design

1. Is one test point per discrete active stage brought out to the connectors?
2. Are functional circuits of low complexity?
3. Are circuits functionally complete without bias networks or loads on another unit?
4. Is each test point adequately buffered or isolated from the main signal path?
5. Is a minimum number of phase or timing measurements required?
6. Does the design avoid external feedback loops?
7. Does the design avoid or compensate for temperature-sensitive components?
8. Does the design allow testing without heatsinks?

FIGURE 5-32   Testability engineering checklist.

Digital design

1. Does the design contain only synchronous logic?
2. Are all clocks of differing phases and frequencies derived from a single master clock?
3. Does the design include data warp-around circuitry at major interfaces?
4. Do all buses have a default value when unselected?
5. Does the design include current limiters to prevent domino effect failures?

Built-in test

1. Can BIT in each item be exercised under control of the test equipment?
2. Does BIT use the building-block approach?
3 Is BIT optimally allocated in hardware, software, and firmware?
4. Is processing or filtering of BIT sensor data performed to minimize BIT false alarms?
5. Does mission software include sufficient hardware error detection capability.
6. Is sufficient memory allocated for confidence test and diagnostic software?

FIGURE 5-32   (*Continued*).

where testability is demonstrated because the whole point of the testability program is to be able to test the system to identify faults. If a known fault cannot be identified using designed-in BIT or ATE, then the testability program has not been successful.

The purpose of testability inputs into the maintainability demonstration is to ensure that the testability requirements of the system design have been met. The demonstration cannot be conducted without inducement of failures into the system for which maintenance can be performed. However, the selection of failures for use is extremely important for the contractor and the customer so that both get an objective evaluation of the overall testability and maintainability of the system. A logical cross section of the types of failures that can be expected to occur after the system is fielded should be used, not just a bunch of faults that were picked arbitrarily at the last minute.

Development and design costs	**Development and production benefits**
Testability program costs	Test generation costs
Testability design costs	Production test costs
Testability analysis costs	Test equipment costs
Testability data costs	Interface device costs
Operation and maintenance costs	**Operation and maintenance benefits**
Unit cost increases due to additional hardware required for BIT and test capabilities	Reduced test and repair costs
	Reduced test and repair time
	Reduced manpower costs
	Reduced training costs
	Reduced spares costs

FIGURE 5-33   Testability cost benefit.

To be a useful tool in verifying that testability and maintainability requirements of the design have been met, the demonstration must be planned and executed in a manner that demonstrates all the fault isolation and detection features of the system. The first issue that must be agreed on before a demonstration can be developed is the parameters of the test. The types of functions to be candidates for maintenance action and the level of maintenance to be simulated must be identified, and the methods for selecting faults must be determined before any detailed planning can take place.

Demonstration Planning

Planning for the testability inputs to the maintainability demonstration should begin early in the design process. The basis for selecting faults for the demonstration should be the same faults that were predicted by the FMECA. Since the testability of the system has been, in theory, designed based on the predicted FMECA failures, it is logical that if any other source were used to select faults, the demonstration would not be representative of the anticipated maintenance actions after the system is fielded. The customer may require a contractor to prepare a fault catalog that contains a number of possible system failures. The catalog may be limited to only the number of critical failures that are predicted for the system, or a finite number of faults may be required. The customer then will use this list to select candidates for the demonstration. The contractor may be allowed to recommend the types of faults to be used; however, final selection must be approved by the customer.

FAILURE MODES, EFFECTS, AND CRITICALITY ANALYSIS (FMECA)

The identification of all the probable ways that parts, assemblies, and the system may fail, the causes for each failure, and the effect that the failure will have on the capability for the system to perform its mission provides a valuable tool for systems engineers and design engineers. A failure modes, effects, and criticality analysis (FMECA) is a complete analysis of each level of the system. This analysis technique combines virtually all related engineering disciplines into a single activity. The key participants in development of the FMECA are listed at Figure 5-34. Using the FMECA, engineers identify each possible failure mode of the system. A failure mode is something that occurs, such as a part failing that causes the system not to function properly. A single part can have several failure modes. Something to be remembered when considering failure modes is that they are actual failures, not symptoms of failures. For example, if you put the key in the ignition of your car, turn it, and nothing happens, that is a failure symptom. The actual failure would be a dead battery, faulty ignition switch, etc. Turning the key and nothing happening is only an indication that a failure has occurred. The FMECA also provides information of failure indicators, or how users know when a failure has occurred.

- Systems engineering
- Design engineering
- Reliability engineering
- Maintainability engineering
- Testability engineering
- Safety engineering
- Supportability engineering

FIGURE 5-34 FMECA development participants.

System ———————————— Date ————————————

Indenture level ———————— Sheet ———— of ————

Reference drawing ———— Compiled by ————————

Mission ———————— Approved by ————————

Identification number	Item/functional identification (nomenclature)	Function	Failure modes and causes	Mission phase/ operational mode	Failure effects			Failure detection mode	Compensating provisions	Severity class	Remarks
					Local effects	Next higher level	End effects				

FIGURE 5-35   FMECA worksheet.

Other information developed by a FMECA includes predictions of the percentage of occurrence of each failure mode for a part, a description of what caused the failure, the effect that the failure will have on the capability of the system to perform its mission, identification of any safety or other type of hazard that the failure will cause, identification of methods required to fault isolate the failure, and corrective action that is required to fix the failure. An example of a FMECA worksheet with a description of the information that is recorded on the worksheet is given in Figure 5-35. Later in this text, the FMECA will be referenced as input information for other supportability engineering activities. The FMECA is an integral part of the process that produces a high quality system that can be supported effectively while minimizing the cost of ownership.

Purpose of FMECA

FMECA has four basic purposes, which are listed at Figure 5-36. Hazard elimination is the primary purpose of FMECA. All potential hazards to equipment or personnel that might happen when a failure occurs are identified and eliminated whenever possible. Where removal through design change is not possible, then cautions and warnings in operation and maintenance manuals may be used to prevent their effect. The second purpose of FMECA is to identify loss-of-function results in the system when it is not mission capable. This event has a direct effect on availability, so every effort is made to develop redundancy or compensating methods to negate or control the loss of function. The third purpose of FMECA is to provide a clear method for determining when a loss of function has occurred and to aid in diagnosing which item has to be repaired. The method of detection entered on the FMECA worksheet develops the troubleshooting logic for the system. Finally, FMECA provides a

- Hazard elimination
- Mission capability
- Diagnostics development
- Support planning

FIGURE 5-36   Purpose of FMECA.

1. Define the system to be analyzed.
2. Construct functional block diagrams.
3. Develop mission phase definitions.
4. Identify all potential item and interface failure modes.
5. Define their effect on the immediate function or item, on the system, and on the mission to be performed.
6. Evaluate each mode's worst potential consequences and assign a severity classification.
7. Identify failure detection methods and compensating provisions.
8. Identify corrective design or other actions required to eliminate the failure or control the risk.
9. Identify effects of corrective actions or other system attributes.

FIGURE 5-37   FMECA development steps.

basis for support planning. For every way the system can fail, there must be a way to return it to an operable condition. FMECA is a technique that joins all participants in the design and development process into a single analysis.

Developing FMECA

FMECA is developed in a very methodical, step-by-step process using the steps listed in Figure 5-37. Its development is the responsibility of systems engineering; however, every engineering discipline is an active participant in completion of the analysis. The first step is to clearly identify the system or equipment to be analyzed. The analysis always should be to the functional system, so the same functional block diagram that was first discussed in Chapter 3 is used again as the basis for FMECA. This is a very important point. There is only one system, so there should be only one functional block diagram that is used by all disciplines. The mission contained in the user requirements study, also discussed in Chapter 3, is also a key input into FMECA because the analysis looks at the effect of failures during every phase of the mission. It should be remembered that FMECA is an analysis of the loss of function and the effect of that loss of function on the system when it is being used to perform its assigned missions. The most valuable use of FMECA is during the early design stages as a significant tool in developing a system that is as safe and mission-capable as possible. FMECA has limited value if prepared after the design is completed.

Achievement of appropriate reliability, maintainability, and testability design characteristics is critical if a system is to be supportable. This chapter has provided the basic philosophies and practices of these engineering disciplines. The statistics discussed in this chapter form goals, objectives, and thresholds for design of a system and then provide a method of assessing the progress toward success. Supportability engineering must work hand in hand with these engineering disciplines to ensure that the final system design attains a reasonable balance among performance, supportability, and cost of ownership.

CONCLUSION

Reliability, maintainability and testability are at the core of supportability engineering. It is obvious from the descriptions provided in this Chapter that these specialty engineering disciplines play a leading role in achieving any system availability requirement. This point will be confirmed in the discussions at Chapter 9, Availability. The FMECA is an analysis technique that draws all these aspects into a single focus during the systems engineering process. While a FMECA does not guarantee a good design, it provides the vision to make any design better. The reliability, maintainability, testability and FMECA techniques presented in this chapter are based on sound engineering practices and must be applied on every program. They form the basis for many of the topics discussed in subsequent chapters in this book and represent the basic concepts that lead eventually to a supportable system.

CHAPTER 6
SUPPORTABILITY CHARACTERISTICS

A system that is successful must achieve a balance among performance, support, and cost of ownership. Systems architecting and system engineering historically focused primarily on performance, with cost of ownership being a secondary concern. Support was viewed as a requirement that could be determined only after the system was in operation. This philosophical view limited the consideration of support issues during system design and acquisition. Today systems architecting and systems engineering must give supportability an equal, and possibly greater, importance. Supportability engineering is responsible for actively participating with the systems architecting and systems engineering processes to attain a balance between performance, support and cost of ownership throughout development and acquisition. The traditional term for this activity is *influencing design decisions* to improve supportability and lower cost of ownership.

THE SUPPORTABILITY ENGINEERING CHALLENGE

Supportability engineering activities are interwoven with those of reliability, maintainability, and testability. It is often difficult to determine where the actions of one discipline stop and another starts. This is the way it should be. Producing a supportable system requires a joint effort by all organizations. They must work together as a team, each team member having a stated role in the overall objectives of the group. As stated previously, supportability is defined as follows:

Supportability   A *prediction or measure* of the characteristics of an item that facilitate the ability to support and sustain its mission capability within a predefined environment and usage profile.

For something to be predicted or measured, it must be intrinsically measurable. This is the core of the historical indifference given to support of a system during development. Given the fact that support was viewed as something that happened after the system was placed in service, it could not be defined or determined adequately until sufficient usage of the system had occurred to see what actually was needed or used to support the system. Chapter 3 described how the need for a system is defined, and Chapter 4 explained how the functional performance requirements for a system are used to develop the final design solution. The user, or buyer, of a system initiates a procurement project by defining the things that the system must do to meet an intended need. The user is very articulate in describing the measurable performance standards that are required for the system to meet the need.

However, while the user also has an understanding of the need to provide adequate support for the system, the description of the need for the system to be supportable historically has not been precise or measurable in the same clear terms as performance. The description of support requirements was stated as aspirations rather than measurable realities. Figure 6-1 lists examples of these aspirations. There is nothing wrong with these issues. They are all valid. The only problem is that none of them is actually measurable. If something is not measurable, it cannot be tested, and therefore, its achievement

- Easy to maintain
- Cost effective to maintain
- Safe to maintain
- Minimum requirements for personnel
- Maximum use of existing personnel
- Minimum requirements for test equipment
- Maximum use of existing tools and support equipment
- Minimum requirements for new resources
- Maximum use of existing facilities
- Maximum use of standard parts
- Quickly prepared for shipment or transport
- Transported by standard modes
- Interface with existing support systems

FIGURE 6-1   Design supportability requirements.

can never be completely realized. The challenge for supportability engineering is to take these aspirations of the past and translate them into measurable requirements that can be input into the systems engineering process. This will allow supportability characteristics to be included as requirements for the system that can be measured as easily as performance characteristics.

Easy to Maintain

The requirement for a system to be "easy" to maintain is common sense. However, what is *easy*? This is the problem. *Easy* suggests that the maintenance of an item should be done in a short period of time, applying fairly simple techniques, with as little chance as possible of doing the maintenance incorrectly and using as few resources as possible. The most important of these three attributes is the speed of completing the maintenance action. Translating *easy* into clearly measurable criteria is the challenge. The activities the user desires can be done easily. As discussed in Chapter 5, maintainability engineering considers this maintenance process in the development of mean time to repair (MTTR). It was pointed out that the MTTR for a system is the weighted-average time to return a system to an operable condition. However, maintainability engineering assumes that the resources necessary to perform maintenance will be available at the location required in sufficient quantities to achieve the predicted maintenance task time. This is where supportability engineering takes over. Each maintenance action identified by maintainability analysis is analyzed to determine the resources necessary for its performance. Figure 6-2 illustrates the seven typical issues that supportability engineering addresses for each maintenance action.

1. *Failure identification.* The maintenance process begins with some detection that a failure has occurred. The detection somehow must be obvious to the user. The design of the system must provide an unambiguous alert of the failure, or some inspection must lead to its identification. The common term for this activity is *troubleshooting.* Failure modes, effects, and criticality analysis (FMECA), discussed in Chapter 5, provides a guide for creating a series of activities that can be performed by the user that should lead to identification that a failure has occurred. Chapter 8 discusses this further. The accuracy of troubleshooting is critical. A false indication may lead to unnecessary maintenance. Any troubleshooting procedures must be simple and straightforward, leading to the actual problem. A troubleshooting procedure should have very few steps, should

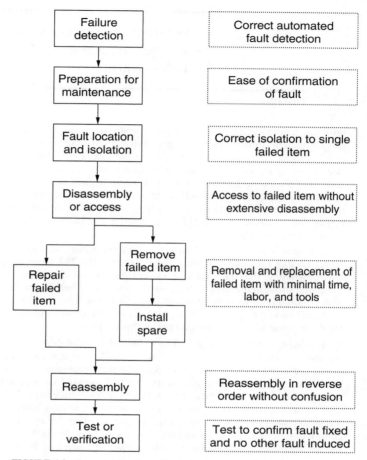

FIGURE 6-2   Supportability issues for maintenance procedure.

provide a clear pathway to avoid ambiguities, and should lead to a single result. Supportability engineering must ensure that all resources required to resolve every possible outcome of the troubleshooting procedure are available. The technical documentation, if required, must be technically accurate, clear in its meaning, and written in a way that the person performing the maintenance action can follow the instructions without error. All tools, support equipment, test equipment, and personnel must be located at the place where the troubleshooting is to be performed.

2. *Fault isolation.* Troubleshooting should lead to the actual item that failed. Some systems may have self-diagnostic capability through built-in test (BIT) or built-in test equipment (BITE). Any self-diagnostics in a system must be accurate. A system having self-diagnostics should indicate clearly that a failure has occurred, and it also should isolate the failure to a single item, which is then repaired or replaced. This may or may not be the actual case. Many self-diagnostic systems only indicate that a failure has occurred and do not identify the failed item. Others may correctly identify that a failure has occurred but will not identify the single item, only groups of items called *ambiguity groups.* In any instance where self-diagnostics are incapable of identifying the failed item, additional testing will be required, or multiple items may be replaced through trial and error

- Self-diagnostics accuracy rate
- Ambiguity resolution
- False alarm rate

FIGURE 6-3 Self-diagnostic concerns.

until the failure eventually is repaired. Additional testing normally requires items of support or test equipment. Finally, there is a potential for self-diagnostics to give a false indication of system failure or erroneously indicate that an item has failed. Either of these events would precipitate performance of an unnecessary maintenance action. Therefore, supportability engineering focuses of three areas listed in Figure 6-3 concerning fault isolation: accuracy of self-diagnostic capability, actions necessary when self-diagnostics cannot isolate a failure to a single failed item, and the potential for self-diagnostics to give a false indication of system failure. Fault isolation methodology is also developed from FMECA information that leads from the functional failure to the physical item that has failed.

3. *Spares, repair rarts and materials.* The results of fault isolation indicate the item that must be repaired or replaced to resolve the system failure. The mean time to repair (MTTR) is calculated assuming that necessary resources will be available to support system maintenance. Supportability engineering is responsible for linking requirements for spares, repair parts, and materials required to support maintenance with the activity that actually develops the resource package that will be provided to support the system. Design characteristics that limit requirements for these items will be discussed later in this chapter.

4. *System design for access.* The system should be designed so that any item that must be repaired or removed can be reached with very few disassembly steps. Figure 6-4 illustrates an electronic assembly. The design of this assembly provides a single cover that, when removed, allows access to any item within the assembly. Figure 6-5 shows a much more complex electronic system where many electronic assemblies have been integrated. Every item within the system is accessible by sliding out sections and then items within each section. This illustration also shows a typical numbering scheme that assigns a unique identifier to every item within the system. The completion of a troubleshooting procedure should indicate the failed item. Then the person performing the maintenance action must be able to relate the output of the troubleshooting procedures to the

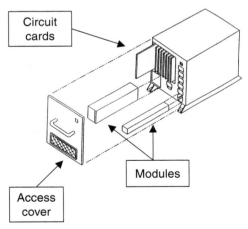

FIGURE 6-4 Design for accessibility: electronic assembly.

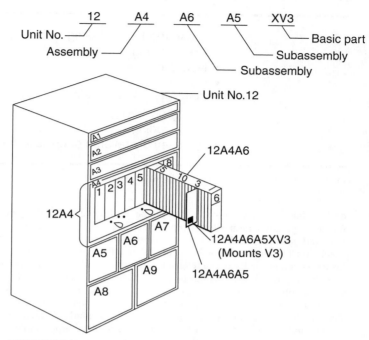

FIGURE 6-5   Design for accessibility: complex system.

physical architecture of the system. The purpose of the numbering system shown in Figure 6-5 is to provide this link. The final issue for accessibility is the physical placement of items within a confined space. For example, it may be necessary to install two electronic assemblies inside a small space in an aircraft. The assemblies cannot be installed side by side owing to space limitations; they must be installed one behind the other. Which one should be installed in front, and which one should be installed in the back? The answer is determined using the mean time between failures (MTBF) of each assembly. Assuming that the assembly with the lower MTBF will fail more often, it goes in front. The assembly with the higher MTBF should be installed in the back. When the assembly in front fails, it can be accessed directly. When the assembly in back fails, the front assembly must be removed first to gain access. By placing items based on reliability potential, the items in a system that fail most often will require fewer steps for access.

5. *Repair or replace.* The decision as to whether an item should be repaired or replaced often is determined using two criteria: time and resources. The MTTR for an item indicates the average time to return the system to an operable condition. The maximum time to repair the system (MAX_{TTR}) is a limitation on time to perform any single maintenance action. The MAX_{TTR} may be the key to determining if an item should be repaired or replaced. Any time for a repair action that nears the MAX_{TTR} is a candidate for further analysis to determine if replacement of the item would require significantly less time. If replacement is technically possible and requires less time, then replacement probably is more desirable than repair. The other criterion for making the repair or replace decision is the quantity of resources necessary for repair compared with the resources required for replacement. There may be instances where removal of the next-higher assembly, the parent of the failed item, is more timely and requires fewer resources. Analysis of each of these options will be discussed as part of level-of-repair analysis (LORA) in Chapter 11.

6. *Disassembly and reassembly.* The design of a system should consider requirements for physical disassembly and reassembly (Figure 6-6). Modularization of the design, where items are

- Clear, readable labeling
- No soldering
- No adhesives
- Captive fasteners
- One-way connectors
- Assembly keying
- Color coding
- Reassembly reverse of disassembly

FIGURE 6-6   Disassembly and reassembly design considerations.

divided into compact segments that can be unplugged and replugged, aids rapid disassembly and reassembly. Each item should have clear, readable labeling to avoid confusion. Items to be removed should not be soldered to other items. They should be joined with connectors. Mechanical items should not be secured with adhesives that make removal more difficult. Captive fasteners should be used wherever possible to avoid losing nuts and washers. Connectors should be one-way so that they cannot be installed backward or upside-down. All items in an assembly should be physically keyed using pins or lugs so that improper reassembly does not occur. Male-female connections should be color-coded where possible to also avoid improper reassembly. Finally, the steps required to reassemble an item should be the exact reverse of disassembly. A design having these characteristics will be much more supportable.

7. *Verification accuracy.* The test required to verify that a maintenance action has been successful should be totally accurate. It is desirable that the same test required for fault detection and fault isolation also be used for repair verification. This eliminates the requirement for an additional test.

Systems whose design characteristics achieve these seven areas are proven to be much more supportable and easy to maintain. The identification and implementation of these design characteristics on a project is through a mutual effort of reliability, maintainability, testability, and supportability engineering.

Cost-Effective to Maintain

This user requirement is the hardest to actually translate into measurable criteria. System effectiveness, as discussed in Chapter 4, was measurable; however, cost-effectiveness, as was shown, is more difficult. The same is true for cost-effective to maintain. The best way to view this aspiration is from the standpoint of the value of having a functional system as compared with having a system that is nonoperational owing to maintenance requirements. The situation must be addressed in terms of the value that the user puts on the system being operational. Rather than embark on a long philosophical discussion at this point, refer to Chapter 9, which presents a detailed explanation of operational availability, and Chapter 10, which discusses cost of ownership.

Safe to Maintain

System safety engineering, discussed in Chapter 7, is responsible for assisting systems engineering and supportability engineering in identifying potential hazards that may cause injury to maintenance or support personnel. This is normally done by analyzing maintenance actions after the design is completed. However, there are several design characteristics that can be included in a system specification to avoid potential hazards. All these tend to be commonsense ideas, but they are worth mentioning at this point.

- Construction materials
- Maintenance materials
- Materials compatibility
- Electrical protection
- Emission protection
- Safety replacement
- Disarming features

FIGURE 6-7   Safety design considerations.

The system should be constructed using materials that are inherently safe. Designs containing radioactive, poisonous, caustic, or other dangerous materials should be avoided unless the technology of the system requires their use. When use of one of more of these materials is necessary, each item containing the material should be clearly marked. There is a tendency to disregard items containing hazardous materials if the operator will not come in contact with them during normal operation of the system. The maintainer is often overlooked (Figure 6-7).

Systems that are constructed with nonhazardous materials still may require hazardous materials for maintenance. Use of some solvents, degreasing compounds, flushing fluids, fuels, lubricants, or solders during maintenance may create a hazardous situation for maintenance personnel. Designs should eliminate the requirements for these items. The biggest concern is when a decision is made to substitute a nonapproved item for one that has been considered originally during design of the system. If this happens, it is typically after the system is placed into service. The change to a nonapproved item most often is due to cost. Someone finds an alternate source of supply that offers a product at a lower price. In an attempt to save money, the nonapproved product is purchased and substituted. The consequences can be potentially lethal.

In certain cases, two materials that are inherently safe can be hazardous when combined. This does not happen very often, but it always should be a concern. This happens most often when construction materials are exposed to maintenance materials that have not been studied for compatibility limitations. This is a very specialized area. Only a materials compatibility expert should be asked to make the final decision on this issue.

Systems must be designed to protect maintenance personnel from electrical shock. A system must be designed with a positive mechanical switch to shut off all electrical power to the system. There are some exceptions to this rule. In certain cases, systems are designed so that maintenance can be performed while the system is still partially operating. In such a situation, one of two characteristics must be inherent in the design: the ability to isolate an item from the power source before starting a maintenance procedure or design of the system to allow removal and replacement of items without exposing maintenance personnel to electrical power sources. Isolation of items within a system requires a series of internal switches in addition using a single master switch. A design that allows removal and replacement of items while the system is operating (called *hot-swapping*) necessitates significant capabilities being designed into the system, where an item is unplugged and plugged back into its power source.

The technology baseline of a system indicates if it presents the potential for hazardous emissions. The emissions may come from the system during operation or may be a by-product during maintenance. Examples of hazardous emissions include gases, vapors, radiation, electrostatic discharge, or liquids. The complete design of a system and all maintenance actions must be analyzed to identify and eliminate or control these emissions.

Some systems that have significant safety issues, especially for operation, use built-in safety devices, such as safety wires or seals. Safety devices installed at the factory typically receive a very high level of inspection to ensure that they have been attached or applied properly. When these same devices are removed for maintenance purposes, the same level of diligence should be used to retain their hazard-free status. However, frequently the design of a system makes reinstallation of safety devices difficult. The safety device may have been applied to an assembly before it was integrated into the system. The assembly is not removed for maintenance, but the safety device must be removed for access. Replacement of the device after maintenance may be extremely difficult. This difficulty is caused by inattention during the design process, not lack of expertise of maintenance personnel.

Systems having an explosive component create additional problems for maintenance. For example, the ejection seat in a combat aircraft contains an explosive device. Any removal of the seat requires disarming. The seat must be removed and the explosive disarmed before any maintenance can be performed on the seat. Occasionally, the seat must be removed to gain access to other items within the cockpit. Again, it must be disarmed. In this case, the aircraft cockpit should be designed so that any item can be accessed for maintenance without removal of the ejection seat.

Minimum Requirements for Personnel

This requirement focuses on limiting the number of people required to perform maintenance. Maintainability engineering analyzes the design of a system to minimize the time required to perform maintenance; however, this does not address the complete issue of the number of people required to perform the maintenance. Determination of the number of people required is based on five issues: (1) how often must each individual maintenance action to be performed, (2) how long it takes to perform the maintenance action, (3) how many people are required to perform the action, (4) the number of systems supported by a maintenance location, and (5) system usage rate. The frequency of performing maintenance actions results from reliability predictions discussed in Chapter 5. The duration of a maintenance action comes from maintainability predictions also discussed in Chapter 5. The number of persons required to perform a maintenance action depends largely on its physical characteristics. Human factors engineering, discussed in Chapter 7, determines the maximum allowable weight for an item and its other physical characteristics that create a requirement for additional people simply to lift or move the item. These issues, combined with the number of systems supported by a maintenance location and the usage rate of the fleet, result in personnel requirements.

Figure 6-8 illustrates how the labor hours required to maintain a fleet can be calculated. However, this does not relate directly to the number of people required to perform maintenance. Here is where

---

$$\frac{\text{Number of systems} \times \text{usage rate}}{\text{MTBF}} = \text{Estimated number of failures}$$

Estimated number of failures $\times$ MTTR $\times$ average persons per task = labor hour requirements

Example: Number of systems supported by a location   90
               Usage rate of each system per year   2500 hours
               MTBF   240 hours
               MTTR   2.25 hours
               Average number of persons per action   1.5

Results: Approximately 3164 labor hours required to maintain the fleet

---

FIGURE 6-8   Labor-hour estimation.

MLH/MA = MTTR × Average persons required per maintenance action

Example: 2.25 × 1.5 = 3.375 labor hours per maintenance action
(using input data from Fig. 6-8)

FIGURE 6-9   Mean labor hours per maintenance action.

the challenge arises. How many people must be present at any one time in order to provide the necessary labor? In theory, if each system failure occurs independently at different times, then only two people are required. In real life, however, several systems may require maintenance at the same time. Is it acceptable for failed systems to wait in line so that each can be repaired in turn? If so, then the answer obviously is that two people are sufficient. Are all the systems required to be mission-capable all the time? These questions are addressed in detail in Chapter 9 in terms of operational availability.

There is another way to approach personnel requirements; this is to create design-to goals that will result in a system that does, in fact, minimize personnel requirements. Two statistics are key to this approach: mean labor hours per operating hour (MLH/OH) and mean labor hours per maintenance action (MLH/MA). Figure 6-9 shows how MLH/MA is calculated, and then Figure 6-10 shows how this statistic is multiplied by the system failure rate to produce the MLH/OH. Both these statistics are design-to goals that can be specified as system requirements.

Maximum Use of Existing Personnel

This user aspiration is closely related to minimizing personnel requirements, but it includes additional considerations. The personnel required for a system are described in terms of the number of people and their capabilities. The number of people was discussed in the preceding paragraph; however, this view is different. The capabilities of people are determined by the knowledge and skills they possess. For example, an old fleet of systems is being replaced by an equal number of new systems. The user desires to use the existing number of people to maintain the new system, and the user also does not want to have to provide any additional knowledge training to these people. The old fleet historically has required 0.01 MLH/OH. If the new design meets this requirement, then the existing number of people should be sufficient. Applying this requirement to the example provided in Figure 6-10 indicates that this design does not meet the requirement.

MLH/OH = System failure rate × Maintenance labor hours per maintenance action

Example: 0.004167 × 3.375 = 0.0141

(Using input data from Figs. 6.8 and 6.9)

FIGURE 6-10   Maintenance labor hours per operating hour.

Minimum Requirements for Test Equipment

This requirement is easier stated in the positive terms of "No test equipment required." Minimum cannot be measured, whereas "No" is clear, concise, and easily measured. This may or may not be possible for a system based on its technology. Test equipment for an electronic system may be eliminated through use of self-diagnostics. Where test equipment is required, the design should be standardized to require only one test set. This design feature is discussed in the following paragraph.

Maximum Use of Existing Tools and Support Equipment

Again, this requirement should be stated in a different way, such as "Use only existing tools and support equipment." The system requirement study described in Chapter 3 contains a section for identification of existing tools and support equipment that will be available to support the new system. This list of items can be provided to design engineers with instructions that only items on the list will be used to support the new system. The final design solution can be measured against this list to determine if the design meets the requirement.

Minimum Requirements for New Resources

Resources to support a system always will be required. In this instance, the user prefers to use what is available rather than developing something that is not currently used for the old system being replaced. Requirements for resources are driven by the technology baseline of the design. Where the new system technology is similar to that of the system being replaced, this requirement may be achievable. However, if the technology is different, then the resources may be significantly different in many ways. Therefore, the user must assess the need for the technology change and weigh the value of increased capability with the cost of providing new resources. Where new resources are required, a vigorous standardization effort should limit the number of these resources needed to support the system.

Maximum Use of Existing Facilities

Facilities represent a long-term investment and are a significant contributor to cost of ownership. Every effort must be made not to create a requirement for new facilities. This is the logic behind this user aspiration. The approach to this issue is identical to the approach described earlier concerning use of existing tools and support equipment. The positive way to state this requirement would be "No new facilities." The system requirement study discussed in Chapter 3 should identify all existing facilities that will be available to support the new system. Specific information on the capacity, technical capabilities, installed equipment, physical dimensions, size and shape of doors, load-bearing strength of the floor, and any other unique aspects should be in the system requirement study. Supportability engineering assists systems engineering, not design engineering, in creating a design solution that does not exceed the limitations of existing facilities. The only justification for additional facilities should be the placement of new system in different locations where there are no existing facilities or a new technology for which the user has no existing facilities. An existing facility that is determined to be inadequate to support the new system should be modified, upgraded, or expanded rather than creating a new facility.

Maximum Use of Standard Parts

The first issue concerning this aspiration is the definition of a *standard part*. Typically, a *standard part* is one that is used commonly by many systems, has multiple sources where it can be purchased, and conforms to a specified material content and physical dimensions and functional capability. There are many established standards published by internationally recognized organizations. Therefore, applicable standards must be selected for the technology of the system being acquired.

Then this requirement can be restated as "Use only parts that conform to selected standards." A list of conforming parts from each standard can be supplied to design engineers. The design engineers then select only parts from these lists to be used in their individual design solution.

Quickly Prepared for Shipment or Transport

This requirement typically applies to systems that may be moved frequently to various locations. The user is trying to say that when it is time to move an item, its design should allow it to be transformed from its operational configuration to its shipping configuration with minimal effort. An example of this feature is two aircraft engines. The first engine, when removed from the aircraft, requires further disassembly preparation before it can be placed in its shipping container. The additional disassembly may consist of removal of fragile items, or subassemblies, that protrude at awkward angles and draining of residual fluids. Holding fixtures must be attached to the engine to keep it upright in the shipping container or attach it to the container. The actual movement of the engine requires a special harness that links to the crane that lifts it into place. These activities may require several hours and necessitate additional tools. The second engine has been designed with no exposed fragile or protruding items. Inside the shipping container is a holding device that the engine can be placed directly into from the aircraft. The only action required to prepare the second engine for shipment is actually bolting it into the container. The second engine has been designed for shipment where the first was not.

Transported by Standard Modes

Here again is the desire to use something "standard." Transport can have several facets: transport of the system to various locations, transport of support materials to the system, transport of failed items to other places for repair, and transport of resources that must accompany the system when it is transported. Each of these situations potentially requires a different type of transport. This is another issue that must start with identification of the transportation modes used for the system being replaced and those used by its support infrastructure. Transportation modes are listed at Figure 6-11. The objective

Land
- Open vehicle
- Closed vehicle
- Containerized
- Towed behind vehicle
- Rail, open car
- Rail, closed car

Air
- Cargo aircraft
- Helicopter, internal
- Helicopter, external

Sea
- Cargo ship, internal
- Cargo ship, external
- Containerized
- Towed behind ship

FIGURE 6-11   Transportation modes.

is to ensure that the final design solution of the system can be transported using the existing transport modes used by the system being replaced unless use of one or more of those modes can be improved to produce a lower cost of ownership.

Interface with Existing Support Systems

A new system is being acquired to replace an older system. The user has an existing support system that sustains all the systems currently in use, not just the system being replaced. It is the user's desire to introduce the new system into this existing support system with no impact on other systems while at the same time optimizing support requirements for the new system. Problems can arise that must be considered.

For example, the system being replaced currently shares a maintenance facility with another system. The new system will not require use of this facility owing to reduced maintenance requirements. This means that the shared portion will become excess when the new system goes into service. The maintenance facility is still required, but the amortization of its annual cost will be totally on the other system and therefore increase its cost of ownership.

As another example, the user may have a fuel-delivery system that is common to all systems in the inventory. The new system must be capable of accepting fuel from this same source. To require a different method would create a significant cost escalation in the infrastructure required to support the new system.

The situation study prepared at the formulative stage of acquisition of the new system should identify all the support systems with which the new system must interface. The only valid justifications for deviating from the existing support infrastructure would be a new technology not currently supported by the infrastructure or a usage profile that the existing infrastructure is incapable of sustaining.

RESTATING THE REQUIREMENTS

This chapter has addressed how the user's aspirations for a supportable system can be translated into understandable and measurable facts. Figure 6-12 shows how the aspirations presented in Figure 6-1 can be restated into clear, measurable design goals. Terms such as *minimum* and *maximum* never should be used because they are subjective and never can be clearly defined. The terms *no* and *must* are very measurable and provide a clear basis for determining achievement. A review of these design goals shows that many may be physically unattainable; however, they do create goals for systems engineering to strive for. A final system design that has characteristics that approach meeting these goals will be very supportable. Supportability engineering is responsible for assisting the user in refining these goals and then participating in systems engineering to strive to attain them.

IMPLEMENTING THE REQUIREMENTS

Implementation of supportability requirements on an actual program must be in concert with the evolving design solution. Timing is more important that degrees of accuracy, especially at the start of a program. The goal of supportability engineering is to lower the cost of system ownership. This concept has been discussed several times already in this book. Interjecting measurable supportability requirements into the systems engineering process is the key to success. The windows of opportunity presented in Chapter 2 are the secrets to success.

Design Program

Window D1. Identification of supportability requirements must start in the first phase of system design (see Figure 2-7). As previously stated, 70 percent of all cost-of-ownership decisions are made during this window of opportunity. All requirements must be contained in the

- Fault detection 100% accuracy
- Error free troubleshooting procedures
- Fault isolation to single failed item 100% accuracy
- BIT/BITE coverage 100% of FMECA failure modes
- Spares for every fault isolation result
- All items accessible without removal of other item
- Modular design
- All replaceable items attached with captive fasteners
- Color coding of all connections
- Reassembly reverse of disassembly
- Repair verification using diagnostic test
- No hazardous materials in system design
- No hazardous materials required for maintenance
- All hazards to maintainer eliminated
- Do not exceed existing number of people
- Do not exceed stated MLH/OH
- Do not exceed stated MLH/MA
- No test equipment
- Use only currently existing tools and support equipment
- Use only currently existing facilities
- Use only parts to a recognized and agreed standard
- No preliminary preparation for transport
- Use current transportation modes
- Use existing support systems unless justified by technology change

FIGURE 6-12   Restated design supportability requirements.

specification that is issued with the request for proposal (RFP). It is extremely difficult to initiate requirements after this point because they will not normally be contained in the succeeding flow of requirements. All the techniques described in this chapter should be performed by the buyer to develop measurable supportability characteristics (see Figure 2-8) to be included in the specification.

Window D2. The contractor's proposed design solution must include methods for ensuring that supportability requirements contained in the specification will be reflected in the final design solution. It is the contractor's responsibility to suggest complementary versions of the techniques contained in this chapter to substantiate how supportability will be included in the systems engineering process for the proposed design solution (see Figure 2-9).

Window D3. Supportability should have equal importance when making the decision as to which contractor to select for design of the system. Measurable supportability characteristics provide the ability to compare each proposed design solution to see which provides the best fit among performance, supportability, and cost of ownership (see Figure 2-10).

Window D4. Systems engineering is responsible for preparing instructions for design engineers and purchasing specialists (see Figure 2-11). All required supportability characteristics must be included in these instructions. Preliminary design review (PDR) should substantiate that every individual, both design engineer and purchasing specialist, has his or her personal requirements for supportability. This is one of the most critical windows for implementing supportability on a design program. It is also one of the most often missed opportunities on programs where the requirements are not included in the design instructions. Any supportability improvement after this point is very difficult because it typically requires redesign, which increases costs and creates schedule delays.

Window D5. Design engineers and purchasing specialists implement supportability characteristics into the evolving design solution during this window (see Figure 2-12). This should be an easy and painless process when measurable requirements have been included in their instructions. Supportability engineers should use the techniques described in this book, especially in this chapter, to assist design engineers and purchasing specialists in meeting their requirements.

Window D6. All measurable supportability characteristics should be a critical issue for successful completion of the functional configuration audit (FCA). This event substantiates that all requirements in the specification have been met. All supportability characteristics must be included as specific, measurable requirements. The prototype or first production system that is tested must demonstrate that all supportability characteristics have been included in the final design solution (see Figure 2-13).

Off-the-Shelf Program

Window O1. Supportability engineering activities during this window are exactly the same as for window D1 because the procurement option is not selected until near the end of this phase. However, the results are stated in terms that can be used to select the system to purchase (see Figure 2-15).

Window O2. The supportability characteristics of the seller's proposed off-the-shelf (OTS) product should be described in specific detail and show how each requirements is met. All the proposed products already have been designed, and most should already be in use by other organizations. The seller should be able to show in real-life terms how his or her proposed product meets all requirements (see Figure 2-16). One thing that must be clearly stated by the seller is the environment of use and the support infrastructure of the original user of the system. Significant differences between that environment of use and the support infrastructure of the original user and that of the OTS purchaser could invalidate any of the supportability results. It is important that the buyer has detailed visibility of the specifics of each current system user. It is highly desirable that every supportability issue be addressed in terms of both the system design and the original user's situation in terms of mission, environment, and support infrastructure.

Window O3. Selection of the appropriate OTS product is the most significant decision affecting supportability and cost of ownership. This is why there are two sets involved to make sure that the right product is purchased. Window O3 (see Figure 2-17) assesses the documentation submitted by the seller to determine which proposed products meet the specified requirements. Every claim of technical conformance should be substantiated with documented proof based on real field usage or extensive testing conducted by the original user. All necessary analyses should be performed to assess the proposed products. Every supportability characteristic described in this chapter should be studied to ensure that the product meets the minimum requirement.

Window O4. The testing performed during window O4 (see Figure 2-18) is to confirm that the proposed products can perform the mission in the user's profile and environment and that the support infrastructure is capable of sustaining the system. Every facet of supportability should be demonstrated during this testing. The audits performed at the completion of physical testing must confirm that all supportability requirements have been met.

Window O5. Final selection of the most appropriate system that provides the best balance among performance, supportability, and cost of ownership is completed during window O5 (see Figure 2-19). An analysis that compares the attributes and characteristics of each possible OTS product is performed to determine the best fit for meeting the user's mission requirements. At the same time, operational availability potential and possible cost of ownership also must be included in the decision-making process.

Supportability engineering develops and applies the characteristics described in this chapter to ensure that the design of the product, whether a new design or off the shelf, meets the user's performance requirements and can be supported in its operational environment. This approach also should result in a system that has the lowest cost of ownership possible while still meeting the user's minimum requirements.

CHAPTER 7
SYSTEM SAFETY AND HUMAN FACTORS ENGINEERING

Supportability of systems is greatly influenced by the efficiency of the people assigned to their operation and maintenance. People are an integral part of the operation and maintenance of systems. Very few items operate in an environment that does not require human contact or involvement with the equipment. Therefore, people must be considered an integral part of the total system. System safety engineering and human factors engineering are responsible for evaluating the equipment design to ensure that it allows human participation to be as safe and efficient as possible. These engineering specialty disciplines address liability, product safety, and health and welfare issues that may arise for any consumer or user of products sold to individuals or organizations.

SYSTEM SAFETY PROGRAM

No matter how good the design of an item of equipment, if it cannot be operated and maintained safely, it is unacceptable. System safety engineering is charged with developing and implementing a system safety program that continually evaluates the evolving equipment design to identify potential safety hazards. As they are identified, hazards are analyzed to determine ways they can be reduced or eliminated through design or procedure changes.

Objectives

The system safety objectives of any program are to influence the design using a systematic analysis and evaluation approach that results in equipment that is as safe as possible to operate and maintain. Figure 7-1 shows the general objectives of system safety. Achievement of these objectives is realized through implementation of the design criteria listed in Figure 7-2. These general design criteria illustrate the areas normally addressed by system safety engineers during equipment design.

Safety Hazards

A *hazard* is a situation that, if not corrected, might result in death, injury, or occupational illness to personnel or damage or loss of equipment. System safety engineers analyze and evaluate the proposed equipment design to identify hazards. These hazards then are classified in terms of severity and probability of occurrence. Figure 7-3 shows the four categories of hazards, and Figure 7-4 provides an example of how the probability of occurrence can be assigned. The combination of severity and probability can be used to develop a hazard-assessment matrix, as illustrated in Figure 7-5, which can be used to prioritize the effort of system safety by identifying the hazards that are most critical to safe equipment operation.

1. Equipment design stresses safety consistent with mission requirements.
2. Hazards associated with the design are identified, evaluated, and eliminated or reduced to an acceptable level.
3. Consideration of historical safety data gathered from other systems.
4. Minimize risk through use of new designs, materials, and production and test techniques.
5. Minimize retrofit by addressing safety concerns in the early phases of acquisition.
6. Recommended design changes are accomplished in a manner that does not increase risk of hazards.

FIGURE 7-1   System safety program objectives.

1. Eliminate hazards through design.
2. Isolate hazardous substances, components, and operations.
3. Locate equipment to reduce hazards to personnel during operation and maintenance.
4. Minimize risks caused by environmental conditions.
5. Design to eliminate or minimize risk created by human error.
6. Consider alternate approaches to eliminate hazards.
7. Provide adequate protection from power sources.
8. Provide warnings and cautions when risks cannot be eliminated.

FIGURE 7-2   System safety design criteria.

Description	Category	Mishap Definition
Catastrophic	I	Death or system loss
Critical	II	Severe injury, minor occupational illness, or major system damage
Marginal	III	Minor injury, minor occupational illness, or minor system damage
Negligible	IV	Less than minor injury, occupational illness, or system damage

FIGURE 7-3   Hazard category table.

Description	Level	Individual Item	Inventory
Frequent	A	Likely to occur frequently	Continuously experienced
Probable	B	Will occur several times in life of an item	Will occur frequently
Occasional	C	Likely to occur sometime in life of and item	Will occur several times
Remote	D	Unlikely but possible to occur in life of an item	Unlikely but can reasonabl be expected to occur
Improbable	E	So unlikely, it can be assumed that occurrence may not be experienced	Unlikely to occur, but possible

FIGURE 7-4 Hazard probability.

Frequency of Occurrence	Hazard Categories			
	I Catastrophic	II Critical	III Marginal	IV Negligible
A. Frequent	1	3	7	13
B. Probable	2	5	9	16
C. Occasional	4	6	11	18
D. Remote	8	10	14	19
E. Improbable	12	15	17	20

Hazard Risk Index	Criteria
1–5	Unacceptable
6–9	Undesirable
10–17	Acceptable with review
18–20	Acceptable without review

FIGURE 7-5 Hazard risk-assessment matrix.

SYSTEM SAFETY ENGINEERING PROCESS

A formal system safety program (SSP) should be established for each acquisition program. The SSP works with other specialty engineering disciplines, such as reliability, maintainability, supportability, and testability, to provide guidance to design engineers and assessment of the evolving design solution. Figure 7-6 provides a list of typical activities performed through the SSP. Typically, a

- System safety program plan
- Safety program reviews and audits
- Hazard tracking and risk resolution
- Preliminary hazard list
- Preliminary hazard analysis
- Safety requirements criteria analysis
- Subsystem hazard analysis
- System hazard analysis
- Operating and support hazard analysis
- Health hazard assessment
- Safety assessment
- Test and evaluation safety
- Safety review of change proposals
- Safety verification
- Safety compliance assessment
- Explosive hazard classification and characteristics data
- Explosive ordnance disposal source data

FIGURE 7-6 System safety program activities.

formal plan is developed to ensure that the system design is as safe as possible to operate and maintain. The purpose of the system safety program plan (SSPP) is to describe in detail the tasks and related activities that system safety engineers and management will accomplish to identify, evaluate, and eliminate equipment hazards. Figure 7-7 lists the safety areas to be addressed in the SSPP. This forms the roadmap for application of safety analysis to the system.

System Safety Program Reviews/Audits

As with previously discussed programs, system safety engineering must be a formal participant at all program reviews. System safety concerns normally are addressed at all preliminary design reviews (PDRs) and critical design reviews (CDRs). Additionally, the developer might require suppliers to convene special program reviews that deal strictly with safety issues. These special reviews might be required to support certification boards in the areas of munitions or aircraft flight readiness.

Hazard Tracking and Risk Resolution

All participating organizations should have in place a formal method for recording the ongoing efforts of safety analyses. This bookkeeping requirement establishes a log for recording potential hazards as they are identified and provides a method for tracking the resolution and disposition of each hazard. The hazard log is an invaluable tool in managing the system safety effort and providing visibility for critical safety problems. An example of a hazard log format is shown in Figure 7-8.

1.0 PROGRAM SCOPE AND OBJECTIVES

Describe the overall scope of the program and how safety engineers will participate in the design process.

2.0 PROGRAM TASKS

Describe how each safety task (MIL-STD 882B) will be accomplished. Describe interfaces between safety and other engineering activities.

3.0 SYSTEM SAFETY ORGANIZATION

Describe the system safety organization. Include organizational charts. Identify the responsibilities and authority of safety personnel. Specifically identify who has the decision-making authority to ensure that safety concerns are addressed. Include name, address, and telephone number of the system safety program manager. Describe the personnel and other resources that will be used to accomplish program tasks.

4.0 SYSTEM SAFETY PROGRAM MILESTONES

Include a milestone chart that tells in detail when each safety task will begin and end. Identify critical activities that must be accomplished in order to produce a hazard-free system. The milestone chart should be keyed to overall program milestones.

5.0 SAFETY REQUIREMENTS AND CRITERIA

Describe how the safety requirements and criteria will be integrated into the equipment design. Provide an explanation of the processes that will be used to channel safety concerns to the responsible design engineering activity.

6.0 HAZARD ANALYSES

Describe the analytical processes that will be used to identify system hazards. Explain how the results will be used to eliminate or reduce hazards. Identify the depth to which each analysis will be used.

7.0 SYSTEM SAFETY DATA

Identify what data will be recorded and how they will be managed. Describe the procedures that will be used to process and disseminate information. Address how deliverable data items will be prepared and submitted.

8.0 SAFETY VERIFICATION

Describe the methods of test, analysis, and inspection that will be used to verify that requirements for system safety are met.

FIGURE 7-7   System safety program plan.

Control number	Part number	Hazard description	Responsible engineer	Date entered	Date closed

FIGURE 7-8   Hazard log worksheet.

Potential Hazard List

All participants may be required to prepare a potential hazard list (see Figure 7.8) early in the acquisition cycle that identifies the possible safety hazards inherent in the proposed equipment design or use environment. This list forms the basis for further system safety activity, and as the design evolves, it can be modified as required to guide the SSP and determine the scope of other SSP tasks.

SAFETY REQUIREMENTS

Achievement of a safe system design requires that concise criteria be developed and implemented throughout system design. The purpose of safety requirements/criteria analysis is to develop these criteria. The preliminary hazard list and the preliminary hazard analysis are used as the basis for this activity. By defining the minimum criteria for elimination or control of hazards, the final design will be as safe a possible. These criteria include

- Development of safety design requirements and guidelines
- Identification of all hazards
- Identification of safety-critical computer software components
- Identification of critical hardware-software interfaces
- Identification of safety-critical software functions

- Ensuring that safety design requirements are incorporated into operation and maintenance manuals (This includes consistent development of cautions and warnings for technical manuals.)
- Labeling of the system to warn users about potential hazards

The safety design criteria should be developed during the first stage of system acquisition. They must be included in the specification for procurement. An example of safety design criteria is provided in Appendix E These criteria are used as the basis for both design of the system and safety hazard analysis as the design evolves into a mature system.

HAZARD ANALYSIS

A system that can be operated and maintained safely must be as hazard-free as possible. To achieve this end, system safety engineers continually assess the evolving system design to identify and eliminate possible hazards. Hazards may be eliminated by design changes, safety devices, labeling, or use of cautions and warnings in operation and maintenance documentation.

Preliminary Hazard Analysis

System safety engineers perform a preliminary analysis of the potential hazards identified on the preliminary hazard list to determine the overall level of hazard that exists in a proposed design. This analysis is used as an input to trade off studies evaluating alternative design and deployment approaches during early acquisition cycle phases and an input to subsequent analysis activities for development of design criteria.

Subsystem Hazard Analysis

Design efforts that integrate major subsystems might require hazard analyses of each segment of the design to determine the potential subsystem hazards. A subsystem hazard analysis is applicable to major weapon system contracts where several pieces of equipment are integrated to complete the system design. The subsystem hazard analysis requires evaluation of hazards related to operation and failures of the subsystem. It also must consider hazards created when two relatively similar subsystems are integrated, and their combination creates additional hazard potential.

System Hazard Analysis

The system hazard analysis evaluates the total system operation and failure modes to determine the effect of these activities on the overall system. Special emphasis is placed on the effect of integrating subsystems and the results of failures in one subsystem that could cause a hazard to other subsystems or degrade the performance of the total system. The system hazard analysis normally is performed by the organization responsible for overall system integration, whereas much of the subsystem hazard analysis may be performed by subcontractors. The results of this analysis are documented in several different ways. This may include a safety hazard analysis report and periodic safety status reports. The assessment of the effect of each hazard is also recorded as part of the failure modes, effects, and criticality analysis (FMECA).

Operating and Support Hazard Analysis

The purpose of this analysis is to focus on supportability concerns. The analysis should identify and evaluate operating and support hazards associated with the environment, personnel, procedures, and related equipment involved with the life cycle of the equipment being designed. Activities evaluated include operation, testing, installation, maintenance, storage, and training. This analysis identifies needed design changes to eliminate potential hazards. This task is intimately related to the human engineering effort described later in this chapter. The operating and support hazard analysis is a direct input into the physical supportability analysis process discussed at Chapter 9. System safety engineers should participate in the maintenance task analysis process to ensure that all maintenance tasks can be accomplished without risk of hazard occurrence. Where hazards do exist, system safety engineers are responsible for the content of cautions or warnings that are included in operation or maintenance technical manuals to protect personnel and equipment during both operation and maintenance.

Health Hazard Assessment

The health hazard assessment identifies design characteristics that pose health hazards to operation and maintenance personnel. The presence of hazardous materials and physical environments (e.g., noise, vibration, and extreme atmospheric conditions) is identified. This task is also related to the human engineering effort and provides input to design changes to correct or reduce inherent risks to personnel. This task also provides information about hazardous materials that are used in the operation or maintenance of a system. This information can be used to project the amounts of hazardous material waste that will be produced as a by-product of maintenance and support processes.

SAFETY ASSESSMENT

A safety assessment is performed near the end of equipment development that tells the user all the unsafe design or operating characteristics of the equipment. The assessment also should contain the controls or procedures required to reduce the risks imposed by these characteristics. It also might identify all the efforts that occurred during design and testing to reduce risks to as low a level as possible. The safety assessment report is essential in conveying to the user the importance of adherence to recommended safety procedures.

Test and Evaluation Safety

Safety is a primary concern when the system being designed is tested or its operation is being evaluated. Each test of the system that occurs during development must include a safety assessment. Normally, a test and evaluation master plan (TEMP) or an integrated test plan (ITP) is prepared that forms the basis for all formal testing. Safety must be included on the appropriate planning documents to highlight its importance. The plan requires that all information related to safety or hazards be considered prior to and during any test or evaluation. This includes testing or evaluation that is conducted in-house by the prime contractor, at other contractor facilities, and at customer centers, laboratories, or ranges. Chapters 13 and 15 discuss these events and provide more detail on when and where safety should be involved.

Safety Review of Engineering Change Proposals, Specification Change Notices, Software Problem Reports, and Requests for Deviation Waivers

The focus of this activity is to ensure that system safety reviews all proposed engineering or specification changes prior to submittal to the design authority for approval. This evaluation is necessary to identify possible impacts on equipment safety that might be caused by the proposed change. Correction of a design problem could cause another more serious problem in the area of safety. A representative of system safety must review and sign all proposed engineering or specification changes when this task is imposed. This review process will be discussed further in Chapter 11.

Safety Verification

As the design matures, the safety status of equipment must be verified to ensure that specification requirements are met. This verification normally is accomplished as a part of the testing process of equipment. System safety participates in the preparation of test requirements and actually conducts tests and demonstrations to verify that the design meets all safety criteria. FMECA should be a primary reference for this verification. The results of testing are used to identify further safety considerations or hazards that must be corrected through design changes, changes to procedures, or additional safety controls.

Safety Compliance Assessment

The final activity that is a culmination of the overall safety engineering activity is to conduct an assessment of the equipment design to verify that it complies with all military, federal, national, and industry codes imposed either contractually or legally. This is a comprehensive report that includes all safety aspects of the equipment. The results of all the tasks of the SSP are included in the assessment. It also includes identification of special design features, procedures, protective equipment, and support resources required to ensure safe use. In certain cases, where the equipment design is inherently low risk with regard to safety, this might be the only system safety task that is required contractually. From a supportability perspective, this task includes identification and evaluation of hazards inherent in the system or that may arise from system-unique interfaces, installation of the system, or during testing, operation, maintenance, or support of the system. The final safety compliance assessment documents specific concerns that must be addressed in logistics technical documentation, including operation and maintenance manuals, special handling and transportation procedures, training courses, and personal protection equipment.

Explosive Hazard Classification and Characteristics Data

When a system being developed contains or uses explosives, explosive materials, or devices containing explosives, safety engineering performs test and creates procedures necessary for explosive hazard classification. The explosive hazard classification of an item dictates specific limitations or requirements for packaging, handling, storage, and transportation. This information is a direct input into development of appropriate packaging, identification of storage issues, and quantification of transportation information.

The desired result of the system safety program is the design, development, and fielding of equipment that is as free of hazards as possible. Through effective implementation of the system safety program, the producer is able to develop equipment that meets the performance and safety specifications established by the government or the commercial user.

HUMAN FACTORS ENGINEERING PROGRAM

All items of equipment, with few exceptions, require human interaction for operation, maintenance, or both. The goal of human factors engineering is to optimize this human-to-machine interface. Various references are available that describe this physical interface. The U.S. government has produced MIL-H-46855B, *Human Engineering Requirements for Military Systems, Equipment and Facilities,* which contains the requirements for including human engineering in the procurement of military equipment. These interfaces are based on measurements of a large civilian population. MIL-H-46855B establishes what a contractor will accomplish to ensure that the human aspect is considered in the development of the equipment being designed. MIL-STD 1472, *Human Engineering Design Criteria for Military Systems, Equipment and Facilities,* and MIL-HDBK 759A, *Human*

Factors Engineering Design for Army Materiel, provide guidelines for the design of military equipment that consider the limitations caused by using humans as part of the system. A formal contractual requirement for human engineering is normally found on only major system design efforts. Similar references are produced and used in the automotive and aviation industries.

The human factors engineering (HFE) process is similar to those required for other disciplines discussed previously, such as reliability and maintainability. The program consists of analyzing human function requirements, participating in the design process to ensure that human factors are addressed adequately, and testing and evaluating the final design to validate its operability.

HFE ANALYSIS

The first step in a human engineering program is to identify and analyze the functions that the equipment is required to perform, as stated in the procurement specification. Particular interest is paid to the equipment operation and maintenance requirements and to the environment in which the equipment will be used. The functions are assimilated in logical flow and processing sequence to identify exactly how the equipment must perform to accomplish its mission.

Once the functional requirements have been identified, then the human-to-machine interfaces that must be accomplished to meet the functional requirements are quantified. The human tasks necessary to operate and maintain the equipment are the human-to-machine interface. These tasks are not developed in a vacuum by human engineers. They are the result of analyses performed through a coordinated effort of reliability, maintainability, maintenance planning, human engineering, and other related disciplines. Remember, the FMECA starts this analysis process, and it continues through the generation of final technical documentation.

When the tasks have been identified initially, they are analyzed to determine which ones are critical, in terms of human engineering, to mission accomplishment and supportability. Critical tasks are those that, if not accomplished as needed to meet a functional requirement, would result in equipment failure to accomplish the mission, create a safety hazard, or significantly reduce equipment reliability. An analysis of critical tasks is performed to identify key human factors data about each task. Figure 7-9 illustrates typical human factors data that are pertinent to the analysis of critical tasks. The results of this analysis are used as input to the design process to ensure that human-to-machine interfaces are optimized.

Design and Development

Human engineering provides input during the design phase based on the results of human engineering analyses. The design is evaluated continually, and changes are recommended to enhance the human-machine interface. Through the use of mock-ups, models, and dynamic simulation, human engineering develops solutions to interface problems and evaluates design alternatives to achieve the best possible human interface situations. Using the design criteria developed initially by the user, human engineering is able to provide quantitative requirements to the design process and the detailed design of work environments. Figure 7-10 illustrates human engineering areas of concern in the design of work environments.

Test and Evaluation

Human engineering test and evaluation of the proposed design are conducted as an integral part of the overall equipment test plan. Normally, no separate tests are conducted for human engineering. The tests are conducted concurrently with engineering design and development testing, performance tests and demonstrations, and maintainability demonstrations. The purpose of this testing is to validate that the equipment can be operated and maintained by typical user in the intended operating environment. Critical performance tasks are given special consideration during testing. All failures that

Item	Description
Environment	Location and condition of work environment
Space	Amount of space required to perform task Amount of space available Body movements required to perform task
Information	Amount of information available to operator Amount of information required to perform task
Time	Amount of time allocated for task completion Frequency of task performance Maximum allowable time for task completion
Resources	Number of personnel required to perform task Tools and other equipment required Instructions and manuals required
Other	Safety hazards Interaction required between crew members Personnel performance limitations Equipment performance limitations

FIGURE 7-9   Human engineering analysis input data.

- Physical person-to-machine interface (physical, aural, visual)
- Physical person-to-person interface (physical, aural, visual)
- Physical comfort of operator/maintenance personnel
- Equipement-handiling requirements (weight, cube)
- Temperature, humidity, etc., to be encountered
- Climate (artic, desert)
- Equipment environment (vibration, noise)
- Usable space avilability
- Effects of special clothing (gloves, NBC, coat)
- Safety and hazard protection
- Mission-related requirements (tactical environmnet)

FIGURE 7-10   Human engineering design considerations.

occur during equipment design and testing are evaluated to determine if human involvement was the cause. This can be user error, maintenance-induced error, or interface deficiencies. The results of these tests are used to recommend design changes, as required, to meet the human performance requirements of the equipment.

The results of the human engineering program are inputs to the development personnel requirements, support requirements, training requirements, and technical manuals. All human engineering activities must be coordinated with those of other disciplines to receive the greatest benefit from this program. The human link in the human-machine interface is often overlooked and therefore is frequently the weak link that significantly reduces the system capability for mission success.

HUMAN ENGINEERING DESIGN CRITERIA

MIL-STD 1472, *Human Engineering Design Criteria for Military Systems, Equipment and Facilities,* and MIL-HDBK 759A, *Human Factors Engineering Design for Army Materiel,* provide the principles and methodology used to integrate the human into the design of equipment to achieve mission success. The goal of this integration is to optimize the effectiveness, efficiency, safety, and reliability of the operation and maintenance of equipment. The quantitative nature of the criteria contained in MIL-STD 1472 and MIL-HDBK 759A aids in analyzing and evaluating the human engineering aspects of equipment design. Subsequent paragraphs provide an overview of typical human engineering design considerations. An example of HFE requirements that might be found in a typical specification is provided in Appendix F.

Controls and Displays.

Most operational human-machine interfaces are done through controls and displays. Controls are devices that the operator uses to regulate or give commands to the equipment. Displays provide the operator with information for decision making and monitoring of the results of manipulation of controls. MIL-STD 1472 describes the selection criteria for the most desirable types of controls to fulfill specific functions. It describes how controls should be grouped and arranged to reduce excessive operator movements; methods for coding, labeling, and identifying controls; and design methods that prevent accidental activation. Displays can be either visual or audio, depending on the type of information being provided to the operator. Visual displays range from simple indicator and warning lights to computer displays. They also include mechanical scales, counters, meters, printers, and plotters. The type of visual display used is commensurate with the content and format of the information available and the precision of display required.

The use of audio displays is limited to warnings and voice communications. Warnings are short, go/no-go messages that require immediate or time-critical responses. They are used to inform the operator of critical situations or to indicate impending danger to personnel, equipment, or both. The integration and compatibility of controls and displays are critical. Collocation of controls with corresponding displays should be a basic design goal. The controls and displays selected should be complementary. The information provided by the display should be sufficient to allow the operator to effectively use the control, and the control should provide sufficient range to receive maximum operation.

Environment

The environment in which equipment will be operated and maintained must be considered in the design process. The basic physical comfort of personnel is directly related to the ability of the equipment to conduct and sustain operations. Environmental considerations can be grouped into three categories: factors that design can control (e.g., interior lighting and temperature), factors that design cannot control (e.g., rain, snow, dust), and inherent design factors (e.g., noise, vibration). Human engineering analyzes the design to identify deficiencies and propose solutions in areas that

- Heating
- Ventilation
- Air conditioning
- Humidity
- Hazardous noise
- Nonhazardous noise
- Vibration
- Lighting

FIGURE 7-11   Environmental design considerations.

can be controlled or are inherent design problems. Additional consideration is given to providing alternatives for coping with and minimizing the environmental problems that cannot be directly controlled by the design. Figure 7-11 indicates environmental design considerations.

Anthropometry

Anthropometry is the study of measurements of the human body. Human engineering uses comparative human body measurements to determine the minimum acceptable sizing and dimensioning for human-to-machine interfaces. DOD-HDBK 743, *Anthropometry of U.S. Military Personnel,* contains data obtained by actual measurement of representative population samples of military personnel. These data are organized into limits of 5 to 95 percent of personnel measured and theoretically provide design limits to accommodate 90 percent of the potential equipment users. As an example, DOD-HDBK 743 shows that only 5 percent of the persons (male and female) were less than 60 inches tall and that 95 percent were 73.1 inches tall or less. Therefore, for human engineering purposes, designs must be able to accommodate personnel who are between 60 and 73.1 inches tall. Figure 7-12 illustrates the anthropometric data for a standing person.

Work Space

The physical dimensions of work space must be adequate for personnel to operate and maintain equipment. Human engineering uses the anthropometric data discussed earlier to evaluate the proposed design of equipment, ensuring that the provisions for work space do not impose unacceptable restrictions or hazards on personnel. Specific areas analyzed include space required for standing and seated operations; standard and special console designs; crew compartments; stairs, ladders, and ramps; entrance and exit through doors and hatches; and surface colors. The key to optimizing work space is to design the space around the person rather than designing the space and then putting the person in it. By designing around the person, costly redesign efforts owing to insufficient work space normally can be avoided. Figure 7-13 shows how work-space dimension requirements are determined.

Maintainability

Human engineering can have a tremendous impact on the maintainability of an item of equipment. The mean time to repair (MTTR) predicted by maintainability engineering is based on the capability of personnel to accomplish maintenance tasks within a specified amount of time. The design must

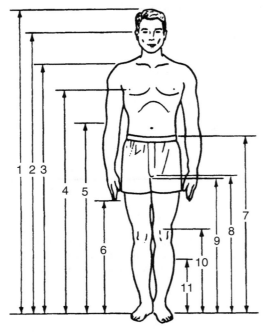

*Same as 12, however, right shoulder is extended as far forward as possible while keeping the back of the left shoulder firmly against the back wall.

FIGURE 7-12   Standing body dimensions.

consider the human interface requirements for performing maintenance. Human engineering evaluates the design to ensure that each maintenance task is as easy to perform as possible. The two keys to achieving maintainability goals, on which human engineering has a direct effect, are accessibility and standardization. If personnel cannot easily access items requiring maintenance, excessive time might be required to perform maintenance. Standardizing fasteners, connectors, and other items that are removed and replaced repeatedly when maintenance is performed standardizes common maintenance tasks, reducing the possibility of human error and increasing proficiency. Specific areas of maintainability interest are accessibility of items to be removed and replaced; lubrication and test points; standardization of fasteners, connectors, and other hardware; design of covers and cases; design for efficient handling; and ease of using tools and test equipment. Significant improvements in the overall equipment MTTR can be achieved through the application of human engineering criteria during equipment design.

Labeling

An area often overlooked is labeling of equipment. Labels are used to identify levels of equipment and state standard procedures and to identify hazards. Labels must be visible and legible, and they also must be durable. The contents, quality, and location of labels that provide directions for operation or maintenance increase the effectiveness of persons using the equipment.

User-Computer Interface

The use of computers is common in most systems and equipment. HFE develops specific human engineering criteria for the user-computer interface. The interface consists of data entry, data display, interactive control, feedback, prompts, error management, data protection, and system response time.

| | PERCENTILE VALUES IN CENTIMETERS | | | | | |
| | 5th PERCENTILE | | | 95th PERCENTILE | | |
	GROUND TROOPS	AVIATORS	WOMEN	GROUND TROOPS	AVIATORS	WOMEN
WEIGHT (Kg)	55.5	60.4	46.4	91.6	96.0	74.5
STANDING BODY DIMENSIONS						
1 STATURE	162.8	164.2	152.4	185.6	187.7	174.1
2 EYE HEIGHT (STANDING)	151.1	152.1	140.9	173.3	175.2	162.2
3 SHOULDER (ACROMIALE) HEIGHT	133.6	133.3	123.0	154.2	154.8	143.7
4 CHEST (NIPPLE) HEIGHT *	117.9	120.8	109.3	136.5	138.5	127.8
5 ELBOW (RADIALE) HEIGHT	101.0	104.8	94.9	117.8	120.0	110.7
6 FINGERTIP (DACTYLION) HEIGHT		61.5			73.2	
7 WAIST HEIGHT	96.6	97.6	93.1	115.2	115.1	110.3
8 CROTCH HEIGHT	76.3	74.7	68.1	91.8	92.0	83.9
9 GLUTEAL FURROW HEIGHT	73.3	74.6	66.4	87.7	88.1	81.0
10 KNEECAP HEIGHT	47.5	46.8	43.8	58.6	57.8	52.5
11 CALF HEIGHT	31.1	30.9	29.0	40.6	39.3	36.6
12 FUNCTIONAL REACH	72.6	73.1	64.0	90.9	87.0	80.4
13 FUNCTIONAL REACH, EXTENDED	84.2	82.3	73.5	101.2	97.3	92.7
	PERCENTILE VALUES IN INCHES					
WEIGHT (lb)	122.4	133.1	102.3	201.9	211.6	164.3
STANDING BODY DIMENSIONS						
1 STATURE	64.1	64.6	60.0	73.1	73.9	68.5
2 EYE HEIGHT (STANDING)	59.5	59.9	55.5	68.2	69.0	63.9
3 SHOULDER (ACROMIALE) HEIGHT	52.6	52.5	48.4	60.7	60.9	56.6
4 CHEST (NIPPLE) HEIGHT *	46.4	47.5	43.0	53.7	54.5	50.3
5 ELBOW (RADIALE) HEIGHT	39.8	41.3	37.4	46.4	47.2	43.6
6 FINGERTIP (DACTYLION) HEIGHT		24.2			28.8	
7 WAIST HEIGHT	38.0	38.4	36.6	45.3	45.3	43.4
8 CROTCH HEIGHT	30.0	29.4	26.8	36.1	36.2	33.0
9 GLUTEAL FURROW HEIGHT	28.8	29.4	26.2	34.5	34.7	31.9
10 KNEECAP HEIGHT	18.7	18.4	17.2	23.1	22.8	20.7
11 CALF HEIGHT	12.2	12.2	11.4	16.0	15.5	14.4
12 FUNCTIONAL REACH	28.6	28.8	25.2	35.8	34.3	31.7
13 FUNCTIONAL REACH, EXTENDED	33.2	32.4	28.9	39.8	38.3	36.5

*BUSTPOINT HEIGHT FOR WOMEN

FIGURE 7-12   (*Continued*).

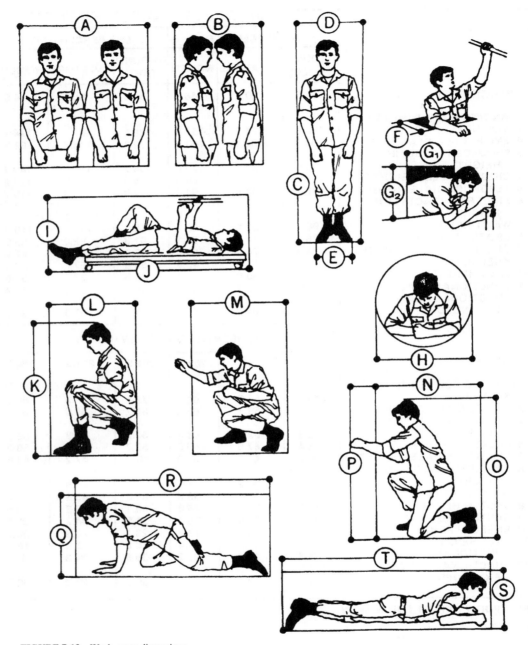

FIGURE 7-13 Work-space dimensions.

	Dimensions (mm)		
	Minimum	Preferred	Arctic Clothed
A. Two men passing abreast	1.06m	1.37m	1.53m
B. Two men passing facing	760	910	910
Catwalk Dimensions			
C. Height	1.60m	1.86m	1.91m
D. Shoulder width	560	610	810
E. Walking width	305	380	380
F. Vertical entry hatch			
Square	459	560	810
Round	560	610	-
G. Horizontal Entry Hatch			
1. Shoulder width	535	610	810
2. Height	380	510	610
H. Crawl through pipe			
Round or square	635	760	810
Supine work space			
I. Height	510	610	660
J. Length	1.86m	1.91m	1.98m
Squatting work space			
K. Height	1.22m	-	1.29m
L. Width	685	910	-
Optimum display area	685	1.09m	-
Optimum control area	485	865	-
Stooping work space			
M. Width	660	1.02m	1.12m
Optimum display area	810	1.22m	-
Optimum control area	610	990	-
Kneeling work space			
N. Width	1.06m	1.22m	1.27m
O. Height	1.42m	-	1.50m
P. Optimum work point		685	-
Optimum display area	510	890	-
Optimum control area	510	890	-
Kneeling crawl space			
Q. Height	785	910	965
R. Length	1.50m	-	1.76m
Prone work or crawl space			
S. Height	430	510	610
T. Length	2.86m	-	-

FIGURE 7-13   (*Continued*).

Each of these areas has significant human factors inherent in the capability of the equipment to accomplish its intended use. The thrust of human engineering effort in optimizing the user-computer interface is to limit the range of inputs and outputs that are possible and that must be processed by the user and to increase the capabilities of the computer to aid the user in performing required operation and maintenance tasks.

HFE TEST, EVALUATION, AND ASSESSMENT

The interface between humans and machines is the critical weak link for most systems. There are various methods for simulation of this interface as the design evolves, but the only sure method to test, evaluate, and assess this interface is through physical demonstration. Every opportunity for physical demonstration of the human-to-machine and human-to-human interfaces required to operate and maintain a system must be used to confirm that the design meets all possible HFE requirements. The assessment processes described in Chapter 13 always must include HFE demonstration. The planning described in Chapter 15 must integrate human involvement for operation and support of any kind throughout the development of the system.

CONCLUSION

The desired result of the system safety and human engineering effort is to influence the equipment design in order to provide environments that maximize personnel productivity and safety and minimize design-induced human limitation and error. Effective integration of system safety and human engineering with other disciplines increases the ability of the contractor to achieve the overall equipment performance and supportability requirements.

CHAPTER 8
RELIABILITY-CENTERED MAINTENANCE

Systems break. This is never desirable, but it does happen. As was discussed in Chapter 5, equipment reliability is a major goal of any development program, but it is impossible to produce a design that does not fail eventually. Since most systems are in fact a consolidation of many related subsystems that all must work together to allow the system to achieve its operational requirements, it is necessary for each subsystem to also achieve as high a level of reliability as possible. A system must perform reliably in order to be available to perform its assigned missions. When a system is inoperative, it cannot perform its mission. Depending on the system's mission, a system failure during the mission not only can be undesirable, but it also can be hazardous or even catastrophic. The purpose of *reliability-centered maintenance* (RCM) is to identify maintenance that can be done on a scheduled basis to avoid unwanted and untimely failures and to improve overall system reliability and therefore system availability. In other words, its purpose is to do something to improve confidence in the reliability of a system or fix an item within a system before it breaks and renders the system inoperable.

A simple example of RCM is periodic servicing of a car. This servicing is done to avoid unwanted failures in the future. RCM is an investment of time and materials to prevent future failures that might happen during system use. There are several other names given to RCM. The most common names include *predictive maintenance* and *condition-based maintenance*. Both these techniques are based on the same concepts as RCM. Development of the RCM process actually began in the commercial aviation industry. Owing to the catastrophic potential of failures during flight of an aircraft, the commercial aviation industry initiated a series of studies several years ago to investigate the possibility of developing a methodological approach to analyzing a system design to identify maintenance that could be done on a scheduled basis that potentially would improve system reliability by avoiding failures.

The results of these studies culminated in a method contained in a document commonly called *MSG-3*, which is the short title for the third report produced by the Aviation Maintenance Study Group. MSG-3 contains a fairly simple logic tree that addressed each potential failure in a system and categorized the failure based on how it would be identified and the potential results if the failure were to occur. MSG-3 is used extensively in the commercial aviation industry and also in the automotive, mining, and petroleum industries. The MSG-3 methodology has been adopted by the military and renamed *reliability-centered maintenance.* The commercial industry uses predictive maintenance extensively as a process of monitoring machinery to detect degraded levels of performance and to take appropriate maintenance actions to avoid complete system shutdown, which can be very expensive.

DEFINITIONS

Preventive Maintenance

Maintenance tasks that are performed on a scheduled periodic basis to prevent future failures are commonly termed *preventive maintenance* tasks. However, we need to draw a distinction between two different types of preventive maintenance tasks: scheduled services and scheduled inspections or removals.

Scheduled Services

Scheduled services are maintenance actions required because of the inherent characteristics of the system. These services include lubrication, alignment, calibration, and other maintenance actions that are required to sustain the operational capability of the system. RCM identifies requirements for scheduled services. These maintenance tasks also may be identified by supportability engineers through discussions with design engineers or maintenance personnel who have experience on similar types of systems. There are also analysis techniques that can aid in identifying requirements for scheduled services.

Scheduled Maintenance

Scheduled maintenance actually consists of corrective maintenance tasks that are done on a scheduled basis. The basis for performance of these tasks is derived by the RCM process. The key to the RCM program is determining when these specific corrective maintenance tasks should be done by performing inspections to determine that a failure is about to occur or to base the performance of the corrective maintenance task on a potential for future failure based on past experience. Therefore, the cornerstones of the RCM program are four different types of scheduled maintenance tasks that trigger performing a corrective maintenance task. These four types of tasks are shown in Figure 8-1.

Scheduled Inspections

There are two types of *scheduled inspections,* on-condition inspections and failure-finding inspections. The basic difference between these two inspections is that *on-condition inspections* look to the future occurrence of a failure, and *failure-finding inspections* look for a failure that has already occurred. A positive result on either of these inspections triggers the performance of a corrective maintenance task just as if a system failure had occurred.

Scheduled inspections:

1. **On-condition task**—A scheduled inspection, test, or measurement to determine whether an item is in, or will remain in, a satisfactory condition until the next scheduled inspection.
2. **Failure-finding task**—A scheduled inspection of a hidden function item to find functional failures that have already occurred but were not evident to the operating crew.

Scheduled removals:

1. **Rework task**—Scheduled removal of units of an item to perform whatever maintenance tasks are necessary to ensure that the item meets its defined condition and performance standards.
2. **Discard task**—Scheduled removal of an item and discard the item or one of its parts at a specified life limit.

FIGURE 8-1   Preventive maintenance tasks.

On-Condition Inspection

The *on-condition inspection* is performed to determine if a failure is about to occur. This means that the item being inspected must wear out through use. The on-condition inspection determines if the item being inspected is about to reach the point through use where it is potentially about to fail through normal wear. An example of this type of inspection is a mechanical item where wear can be gauged and evaluated based on projected use. Each time the item is inspected, the criterion is that it will continue to perform as required until the next periodic inspection. If the item fails the inspection criterion, a corrective maintenance task is performed to replace the item just as if it had failed. On-condition inspection must have an established measurable definition of goodness in order to be effective.

Condition Monitoring

On-condition inspection also may be performed as *condition monitoring*. Condition monitoring is performed typically by a special category of testing equipment that may be an integral part of the system design. A modern naval ship has a condition-monitoring system that continuously monitors critical machinery, such as the propeller shaft, gearbox, engine, and main bearing, for changes in vibration. A change in vibration may indicate that the item being monitored has experienced sufficient degradation to be close to failure, or it may require additional lubrication. The condition-monitoring system consists of a series of vibration sensors. These sensors report the change in vibration electronically, which triggers personnel to investigate further. Condition monitoring also can be performed by additional equipment included in an aircraft design, for example, a health and usage monitoring system (HUMS). The HUMS performs continuous condition monitoring while the aircraft is in flight. Critical indicators identified by the HUMS are reported to the crew for immediate attention. Noncritical indicators are logged and downloaded by maintenance personnel after each flight to identify required maintenance or inspection prior to dispatching the aircraft for the next flight.

Failure-Finding Inspection

A *failure-finding inspection* looks for a failure that has happened since the last inspection. This inspection is the exact opposite in philosophy from the on-condition inspection. An example of a failure-finding inspection is inspection for corrosion, structural cracks, or other minor failures that have not yet caused the system to fail but will lead to a system failure if not corrected. Cracks in the wings of an aircraft, cracks in the blades of a turbine engine, or leaks in the hull of a ship are typical failures that are the subject of a failure-finding inspection.

Scheduled Removals

The second half of the preventive maintenance program consists of *scheduled removals*. These actions are performed on a regular basis based on some measured time or usage of the system. The removals consist of *removal for rework* or *removal for discard*. Scheduled removals are used when neither the on-condition inspection nor the failure-finding inspection has been deemed acceptable to ensure future operability of the system.

Rework Removal

Some items are known to be more prone to potential failure as they are used. These types of items tend to be those which are a major contributor to the system operation. When these items reach a specified amount of usage, they are removed from the system and refurbished or overhauled and then reused. An example of this is aircraft engines. Engines wearout is typically a time-based method measured in hours of operation. When an aircraft engine reaches a predetermined number of hours of operation, it is removed and overhauled. This gives a higher confidence in the reliability of the engine.

Discard Removal

In some cases, items are *removed* at predetermined intervals owing to normal wear, but the item is simply *discarded* rather than refurbishing it and reusing it. Discard of the item may be based on economic considerations, where it has been determined to be cheaper to purchase a new item rather than repair the item that has been removed. Or discard of an item may be based on its physical properties. Some items deteriorate over time regardless of the amount of use they have experienced; batteries, rubber products, and other time-sensitive items fall into this category. Some organizations categorize these as "hard-time tasks." For example, every 18 months the explosive charge in a combat aircraft ejection seat is replaced to ensure that it is fully functional.

Predictive Maintenance

Predictive maintenance is the name given to the activity of identifying equipment that requires maintenance before significant degradation or total shutdown occurs. This requires determination of the point where the degradation starts, which typically is long before there is any significant evidence of the failure potential. To be successful, predictive maintenance must be applied to items that have a clear and consistent wear rate that can be related directly to system usage. This type of maintenance may be performed before there is any indication to the operator of a maintenance requirement.

Condition-Based Maintenance

Condition-based maintenance is similar to predictive maintenance, but it is more applicable to items that do not have a consistent wear rate or a fluctuating usage profile. Condition-based maintenance responds to triggers or preestablished thresholds.

SCHEDULED MAINTENANCE TASK COMBINATIONS

As shown in the analysis process described later in this chapter, there are some instances where a combination of tasks may be used to determine when a maintenance task should be performed. The maintenance personnel who inspect the tires of an aircraft use a gauge to determine tread depth of a tire to identify unusual wear and also remove tires according to numbers of landings. This is a combination of an on-condition task (using the gauge) and a discard removal (based on landings or taxi distance).

RCM ANALYSIS PROCESS

The RCM analysis process considers the significant items that comprise a piece of equipment. It uses information generated by failure modes, elements, and criticality analysis (FMECA) to identify the items that are most critical to the reliability of the equipment and where a failure would have the greatest effect on availability. Figure 8-2 illustrates the analysis process and shows how both design

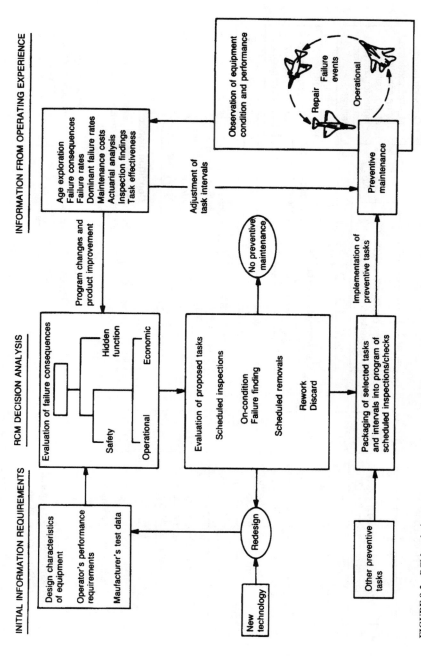

FIGURE 8-2 RCM analysis process.

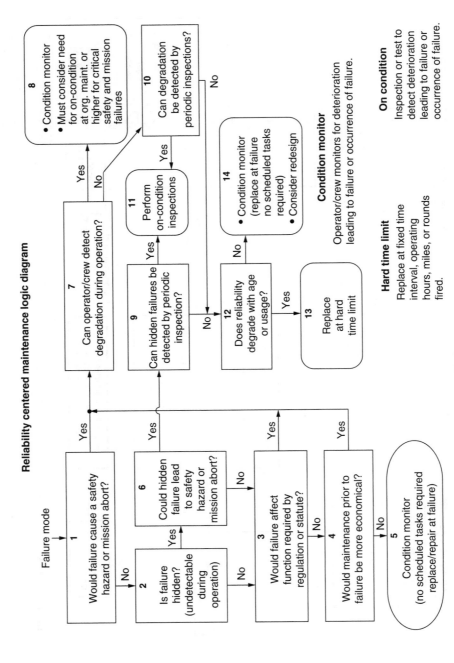

FIGURE 8-3 RCM decision diagram (example from DA Pam 750–40).

and field data are used during the analysis process to identify preventive maintenance tasks. The analysis has proven to be effective in planning preventive maintenance programs for new systems during the development process and when upgrading an existing maintenance program for a system that has experienced significant field use.

Decision Logic

The key to the RCM analysis is the RCM decision logic shown in Figure 8-3. Using this decision tree as a guide, a complete analysis of each significant item can be conducted. The RCM logic is applied to each failure mode listed on the FMECA worksheet, which was discussed at Chapter 5. The results of the analysis provide a clear decision as to what preventive maintenance tasks should be developed to support a system. As shown in the decision diagram, there is a step-by-step process consisting of 16 yes-or-no questions that lead the analyst to decide which type of task, if any, is required. The decision diagram presented at Figure 8-3 is one of the more simplistic versions. There are several variations on this basic diagram contained in different RCM standards. Figure 8-4 provides one of the more detailed examples of an RCM decision tree. Following the logic of the analysis tree that leads to an RCM decision is the summary output for recording the results of the analysis process. This iterative process is used to evaluate each maintenance-significant item to determine if a preventive maintenance task is warranted. The consolidated results of the RCM analysis process form the preventive maintenance program for the system. Note that the decisions are divided into four areas: safety, operational, economic, and hidden failure detection. Each area is related to the activities of several disciplines that should be involved in the RCM analysis process.

Task Frequency

The RCM logic diagram results in identification of scheduled maintenance tasks; however, there is another issue that must follow: determination of the frequency that the tasks should be performed. It should be pointed out that the majority of tasks identified will be for items that deteriorate or degrade over time at a fairly constant rate. Also, the deterioration or degradation must be observable and, hopefully, measurable. Figure 8-5 illustrates the concept of determining task frequency. An item should be in a fully functional condition for some period of time. Eventually, a defect occurs that will lead to a functional failure. The amount of time between onset of the defect and the functional failure provides the ability to determine task frequency. In theory, the on-condition or failure-finding inspection should be performed so that there are two opportunities to detect the defect before it results in functional failure. Therefore, the period of time between defect and functional failure is divided by 2, and this becomes the task frequency. As illustrated at Figure 8-5, the rate of deterioration is the key to determining the frequency. The source for this information should be the design engineer of the item or a specialist who has expert data about the item, such as a materials engineer or mechanical engineer. Where no empirical information is available, a series of stress tests may be appropriate. Also, the environment of use, discussed in Chapter 3, has a significant effect on rate of deterioration and must be considered when determining rate of deterioration.

Cost-Effective Tasks

Performance of scheduled maintenance tasks requires resources such as time, personnel, and possibly, support equipment. This requirement extends potentially to additional training and technical documentation. Therefore, before the final decision is made to perform a task, a cost-benefit analysis must be performed. Figure 8-6 illustrates the concept of this analysis. The analysis compares the cost of performing the scheduled maintenance task with the cost of not performing the task. Determination of the cost of performing the task is fairly easy. This will be discussed further in Chapter 10. It is more

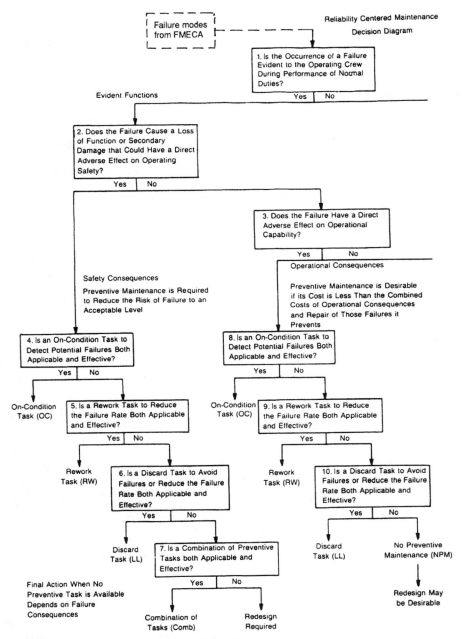

FIGURE 8-4 RCM decision diagram (example from MIL HDBK 2173).

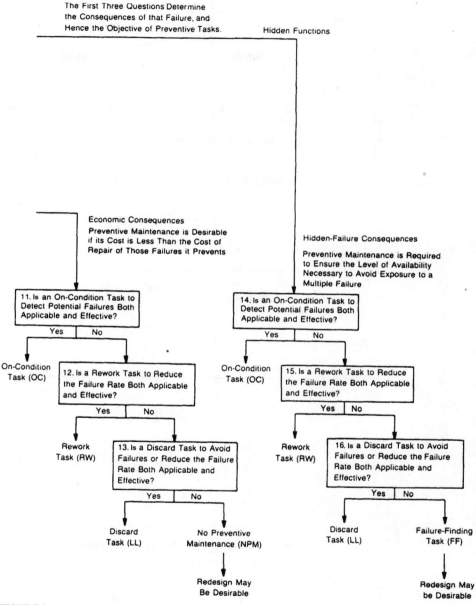

The First Three Questions Determine
the Consequences of that Failure, and
Hence the Objective of Preventive Tasks.

Hidden Functions

Economic Consequences

Preventive Maintenance is Desirable
if its Cost is Less Than the Cost of
Repair of Those Failures it Prevents

Hidden-Failure Consequences

Preventive Maintenance is Required
to Ensure the Level of Availability
Necessary to Avoid Exposure to a
Multiple Failure

11. Is an On-Condition Task to
Detect Potential Failures Both
Applicable and Effective?

14. Is an On-Condition Task to
Detect Potential Failures Both
Applicable and Effective?

Yes   No

Yes   No

On-Condition
Task (OC)

On-Condition
Task (OC)

12. Is a Rework Task to Reduce
the Failure Rate Both Applicable
and Effective?

15. Is a Rework Task to Reduce
the Failure Rate Both Applicable
and Effective?

Yes   No

Yes   No

Rework
Task (RW)

Rework
Task (RW)

13. Is a Discard Task to Avoid
Failures or Reduce the Failure
Rate Both Applicable and
Effective?

16. Is a Discard Task to Avoid
Failures or Reduce the Failure
Rate Both Applicable and
Effective?

Yes   No

Yes   No

Discard
Task (LL)

No Preventive
Maintenance (NPM)

Discard
Task (LL)

Failure-Finding
Task (FF)

Redesign May
Be Desirable

Redesign May
be Desirable

FIGURE 8-4   (*Continued*).

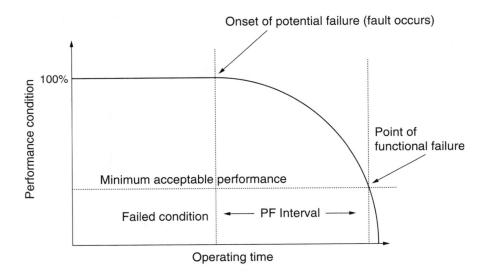

$$\text{Task frequency} = \frac{\text{PF Interval}}{2}$$

FIGURE 8-5 Task frequency calculation.

$$\text{CBR} = \frac{C_{PM}}{C_{NPM}}$$

(CBR < 1 indicates task is cost effective)

$$C_{PM} = C_I + C_{PF}$$

where C_I = cost of one preventive task

        C_{PF} = cost of correcting one potential failure

           = (DMMH for repair × labor cost) + (materials cost)

$$C_{NPM} = C_{CM} + C_{OPC}$$

where C_{CM} = cost of one corrective task

        C_{OPC} = cost of lost operational time

$$= \frac{\text{DMMH for repair} \times \text{system acquisition cost}}{\text{Total operating hours}}$$

FIGURE 8-6 Cost-benefit ratio.

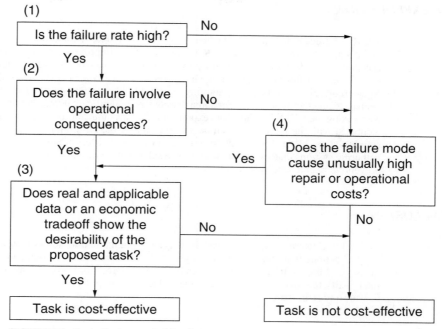

FIGURE 8-7   Cost-effectiveness decision diagram.

difficult to estimate the cost of not performing the task. The estimate must consider the cost of performing the eventual corrective maintenance task when the functional failure does occur plus the loss of use of the system. Any scheduled maintenance task that is designed to prevent a safety hazard or to prevent catastrophic loss of the system should not be subjected to a cost-benefit analysis: Do the task! If safety is not an issue, then a cost-effectiveness decision normally is performed to determine if it makes sense to perform the preventive task. Figure 8-7 shows the typical decision process to make this final decision.

Maintenance Schedule

The RCM analysis process results in a list of schedule maintenance tasks. The task frequency determination will produce the interval for task performance. Typically, the task frequencies will be random periods of time or events. It is normal for the tasks to be consolidated into reasonable time increment groups. For example, three tasks may have frequencies of 28, 32, and 36 days, respectively. Rather than performing these individually at the exact time increment, it tends to be more cost-effective to group them into a single activity consisting of all three tasks that is performed every 30 days. Moving the 28-day task to 30 days may create some risk because the interval would be greater than the time calculated earlier to ensure two opportunities to detect the degradation. Detailed analysis of the deterioration rate and the environmental factors should be conducted to assess the risk of increasing the frequency. This is another analysis where the supportability engineer should have significant involvement.

AGE EXPLORATION

The RCM process does not cease when the system is fielded. A continuing analysis of the preventive maintenance program is conducted to identify areas for improvement. This extension of the process is called *age exploration*. Since many preventive maintenance program are planned and implemented initially using insufficient data, this technique is necessary to achieve the maximum benefits of RCM. Age exploration is a systematic gathering of actual field operation data that are used to refine the preventive maintenance program. For some maintenance-significant items, there may be no clear answer when using the RCM diagram. Age exploration provides that ability to resolve any risks that may be identified with the items. This can be an extension of the failure reporting, analysis and corrective action system (FRACAS) process discussed in Chapter 5. Most organizations have established data-gathering systems that are used to accumulate sufficient field data to thoroughly review and refine the preventive maintenance program.

CONCLUSION

Reliability Centered Maintenance analysis is a proven method for identification of maintenance and support actions that can be done to improve system availability by reducing the severity and frequency of failures. It does require expenditure of resources for implementation, but the results more than justify the investment. This valuable technique is used by any industry where product liability or warranty is an issue.

CHAPTER 9
AVAILABILITY

The product selected by or provided to an owner or organization must meet its intended use or purpose. As has been discussed in previous chapters, the product must have the performance and supportability characteristics necessary for it to achieve a predetermined need. From the viewpoint of the owner, the goodness of the product is measured ultimately in being able to use the product when the need arises. It must be available for use when needed. Before starting a detailed investigation of these issues, a definition is necessary.

Availability   The probability that an item is in an operable and committable state when called for at an unknown (random) time.

CONCEPT OF AVAILABILITY

Having a system for the sheer act of possessing it gives no value to the owner. Value in possessing a system comes through the ability to actually use the system. Therefore, the ability to use a system becomes the final gauge of its value to the owner. The most common term for this value of a system is *availability,* or the ability to use the system when it is required. This concept can be compared to owning an automobile. Just having an automobile does not give sufficient value to the owner unless the automobile provides some form of transportation on demand. For example, if a person owns an automobile, but the automobile is inoperative owing to some failure, then the owner cannot use the automobile. Also, if there is no fuel or materials to make a repair or other basic essentials, the automobile is totally useless for transportation; the automobile is not available for use. It is only when the automobile is available for use that the owner has some sense of value in ownership. The same holds true for any item. Therefore, the concept of availability has been developed as a primary gauge of the value of any product.

Availability only applies at the system level, or the level where the user requires a complete item to perform a mission. Availability is applicable to a car, an airplane, a tank, etc. Availability is not applicable to individual parts or subsystems that cannot perform a mission alone. When applied to a single system, availability is stated as "What is the probability that the system will be fully mission-capable when I need to use it at some random time in the future?" When applied to a fleet of systems, such as aircraft or tanks, availability is stated as "What percent of my fleet will be fully mission-capable at any time in the future?"

Measures of Availability

Availability can be predicted, and it can be measured. Early in the development of a system, availability requirements are created based on the intended mission, usage rate, and environment of use. These form the basis for availability goals or projections that guide the system engineering process. As the preliminary design of a product evolves, it is common to develop an availability prediction for the final design when it is fielded or placed into service. The combination of reliability statistics such as the mean time between failures (MTBF) and maintainability statistics such as the mean time to repair (MTTR) predictions can be used as an early input to predict the amount of time that the system will be available for use when fielded. As the design matures, then other factors are added to the equations for availability prediction. Availability predictions are useful in conducting tradeoff analyses of different maintenance concepts and system design concepts.

Availability Predictions

There are three different versions of availability predictions—inherent availability, achieved availability, and operational availability—each having slightly different inputs and purposes. These analyses cannot be performed with any accuracy until the necessary inputs are available from reliability and maintainability; however, assumptions or allocations can be used and updated until the actual results are in from reliability and maintainability predictions. The measure of predicted system availability also can be used when evaluating different cost factors for support concepts. In many cases it has been shown that reaching an acceptable level of availability costs the same as increasing the availability level by only a few percentage points. For example, the basic support package for a system may cost $10,000. This package is estimated to provide support to repair 90 percent of the predicted system failures that may occur during a typical mission. Increasing the scope of the support package so that 95 percent of the predicted failures could be repaired may cost an additional $10,000 for more spares or other items. The question that must be answered is whether or not the 5 percent increase in availability justifies the 100 percent increase in support-package costs. Under some circumstances, the costs may be justified; however, in most cases, it may be more reasonable to go with the reasonable availability for a reasonable cost. This is a classic tradeoff decision, balancing the relationship between availability and cost of ownership.

A word of warning: Availability only applies realistically to a system because it is the system in total that is required by the user/owner to meet a need. The following availability statistics can be calculated easily for any item such as a fuel pump or circuit card; however, this is not realistic because these items alone cannot perform the intended mission.

INHERENT AVAILABILITY (A_I)

The prediction of the *inherent availability* (A_I) of a system design is the simplest of the three predictions for availability. The purpose of this prediction is to determine the net percentage of time that the system theoretically should be available for its intended use. As shown in Figure 9-1, the only input data required to perform this prediction are the MTBF and organizational-level MTTR, consisting of only corrective maintenance tasks, for the system. The underlying assumptions for this prediction are that system failure will occur at a fairly consistent interval, and whenever a failure occurs, the necessary support (i.e., spares, support and test equipment, personnel, etc.) will be available to perform whatever maintenance is required to return the system to an operational state in the predicted amount of time. Figure 9-1 demonstrates that the inherent availability (A_I) for a system is normally close to 100 percent. This calculation is an estimate of the percentage of time that a system will be available, assuming a near-perfect situation. The system will fail when it is predicted to fail, and the resources necessary to restore it to an operative condition always will be at the point of need in the appropriate quantities. The inherent availability calculation does provide a starting point to compare the new system design with other existing systems to determine the adequacy of the maintenance concept and of the system's inherent reliability. The A_I figure normally is developed early in the design process and is used as a gauge of availability potential. There are some significant shortcomings with this prediction. It assumes a prefect world where a system always will fail exactly when predicted to fail and that all resources required will be available at the proper location to complete a repair exactly within the amount of time that was predicted. Since this never happens, A_I is only a useful gauge early in a development program before the details of a system are known.

$$A_I = \frac{\text{MTBF}}{\text{MTBF} + \text{MTTR}}$$

- -

Example: MTBF = 100 hours
           MTTR =   1 hour

$$A_I = \frac{100}{100 + 1} = 0.99 = 99\%$$

FIGURE 9-1   Inherent availability formula.

The inherent availability formula also can be used to develop or compare the relationship between reliability and maintainability goals for a project. For example, Chapter 5 discussed the development of early requirements for MTBF and MTTR. Combining these statistics into the inherent availability formula allows the resulting combination to be evaluated as to the potential to meet the overall use requirements. Figure 9-2 shows how the A_I formula can be used to establish an MTTR goal by working

Mission time   = 12 hours
Mission success required   = 99%
Availability requirement   = 99.9%

Step 1–Calculate reliability goal using $R(t) = e^{-\lambda t}$ (refer to Chapter 4

   Results: Reliability goal = 1205 hours

Step 2–Calculate maintainability goal using A_I formula

$$A_I \text{ goal } 0.999 = \frac{1205}{1205 + X(\text{MTTR})}$$

or:

MTTR goal X = (MTBF/A_I) – MTBF = (1205/0.999) – 1205 = 1.2 hours

(Note: All figures have been rounded for simplicity and are approximations.)

FIGURE 9-2   Calculating an MTTR design goal.

from a starting point of mission time and availability requirements. As can be seen in this example, the relationship between these three figures of merit form an important basis for eventual supportability of the system when it goes into operation.

ACHIEVED AVAILABILITY (A_A)

The calculation of *achieved availability* (A_A) goes a step beyond that of A_I because this prediction includes the factor *mean time between maintenance* (MTBM) rather than MTBF. MTBM makes allowances for periods when the system will not be available owing to preventative maintenance activities or when maintenance will not be performed but a failure may have occurred. Figure 9-3 illustrates how MTBM is calculated. Also used in the calculation of A_A are the *mean corrective maintenance time* (M_{CMT}) and *mean preventative maintenance time* (M_{PMT}) for the system. M_{CMT} is the average time required to perform a repair action when the system fails, and M_{PMT} is the average time required to perform a preventive maintenance action on the system at a specified interval.

Figure 9-4 provides the formula for calculating system achieved availability. The achieved availability prediction is somewhat more realistic that the inherent availability prediction owing to the fact that is does include all maintenance actions, both corrective and preventive, that will be required to maintain the system. However, the A_A number still does not reflect the real-world requirements for maintenance that the system will experience when it becomes operational. One thing to point out is that if a system does not require any preventative maintenance, then the predictions for A_A and A_I may be almost identical. As with inherent availability, achieved availability has severe real-world limitations simply because of the statistical nature of the calculation. This statistic is best used in the early stages of system development to identify significant areas where availability may be degraded owing to design issues. It is most useful on systems that have a significant requirement for preventive maintenance actions. These systems tend to be large or complex mechanical systems. Using A_A as a design guide during the early development of a system allow requirements for preventive maintenance to be evaluated in terms of their negative affect on availability before the design actually is finalized. Where appropriate, design characteristics that create the requirement for preventive maintenance can

$$MTBM = \frac{1}{(1/MTBF) + (1/MTBPM)}$$

(MTBPM = mean time between preventive maintenance)

FIGURE 9-3   MTBM formula.

$$A_A = \frac{MTBM}{MTBM + M_{CMT} + M_{PMT}}$$

MTBM = mean time between maintenance
M_{CMT} = mean corrective maintenance time
M_{PMT} = mean preventive maintenance time

FIGURE 9-4   Achieved availability formula.

be assessed to determine if a design change would be more cost-effective than performing preventive maintenance over the life of the system.

OPERATIONAL AVAILABILITY (A_O)

The actual gauge of the availability of a system is what percentage of the time under actual operating conditions it is available to perform its mission. This is termed *operational availability* (A_O). Calculation of the actual A_O for a system is accomplished on a routine basis by commercial organizations and by the military services as an indicator of how well the systems actually are providing the ability to respond to consumer requirements or to perform intended missions and how well support organizations are sustaining the systems. Detailed recordkeeping is necessary to support calculation of the A_O for a system. The basic formula for calculation of A_O is provided in Figure 9-5. This formula shows that A_O is calculated for a specified period of time when the system could be called on to perform its mission.

Normally, A_O is calculated on an annual basis to project the percentage of time over a year that the system should be available to respond to a requirement, or it can be stated as the probability that the system will be mission-capable at any random time. The key difference between the formula for calculating A_O and those for other availability measures is that all time over a specific interval is included in the measurement whether the time is operational, maintenance, or down time awaiting maintenance, spares, or other resources required to return the system to an operable condition. This is the real system availability and shows the actual time, on average, that a system can be depended on to be capable of performing its mission or the system being capable of starting a mission at a random time of demand. The basic concept of A_O calculated in Figure 9-5 can be restated as shown at Figure 9-6 to highlight the impact of non-mission-capable time (NMCT) on overall system operational availability. This formula shows that NMCT is the key ingredient in determining A_O.

Operational availability is used to predict the amount of time that an item will be usable over a stated time period. A simple example of how this works can be applied to a person's car. First, there are 8760 hours (365 days × 24 hours) in a year. If a person's car never required repair, never required servicing, and never had to wait for support, then it would be available for use anytime, day or night, for the entire year. Figure 9-7 illustrates this concept. However, this is not realistic because cars occasionally do require repair and normally are serviced a few times each year and sometimes must wait for assistance. Figure 9-8 shows operational availability calculated for a car that required one repair over a year that required 5 days (120 hours). The car also was left for servicing four times, which required 16 hours each time. And the car broke on the highway seven times, and the driver had to wait 3 hours for roadside assistance each time. The operational availability of the car now can be calculated more realistically using these data, as shown at Figure 9-8. The results of these calculations show that in actual terms, the car was available 97.65 percent of the year, which means that in the future, there is a 97.65 probability that the car will be mission-capable in the future, assuming that there is no more degradation in reliability and no lack of required support.

NMCT is the focus of supportability engineering. Chapter 5 presented the concepts of reliability, maintainability, and testability. These are all considered in determining NMCT; however, there is another issue that has been ignored by the previous availability statistics. This issue is the responsiveness of the support infrastructure to provide resources to sustain the system. Figure 9-9 shows that

$$A_O = \frac{\text{mission-capable time}}{\text{total measured time}}$$

FIGURE 9-5   Concept of operational availability.

$$A_O = \frac{\text{MCT}}{\text{total time}}$$

$$\text{Total time} = \text{MCT} + \text{NMCT}$$

Restated:

$$A_O = \frac{\text{MCT}}{\text{MCT} + \text{NMCT}}$$

$$\text{MCT} = \text{total time} - \text{NMCT}$$

Therefore:

$$A_O = \frac{\text{total time} - \text{NMCT}}{\text{total time}}$$

FIGURE 9-6   Operational availability formula restated.

$$A_O = \frac{8760 \text{ hours per year} - [\,(\text{no repair time}) + (\text{no service time}) + (\text{no waiting time})\,]}{8760 \text{ hours per year}} = 100\%$$

FIGURE 9-7   Basic availability prediction with no NMCT.

$$A_O = \frac{8760 \text{ hours per year} - [\,\overset{(\text{repair time})}{(120 \text{ hours})} + \overset{(\text{service time})}{(64 \text{ hours})} + \overset{(\text{waiting time})}{(21 \text{ hours})}\,]}{8760 \text{ hours per year}} = 97.65\%$$

FIGURE 9-8   Basic availability prediction with NMCT.

$$\text{NCMT} = \text{TCM} + \text{TPM} + \text{ALDT}$$

FIGURE 9-9   NMCT formula.

NMCT consists of three parts: total corrective maintenance time (TCM), total preventive maintenance time (TPM), and administrative and logistics delay time (ALDT). Each of these factors is calculated separately and then combined in the A_O formula to determine their total impact on availability.

TCM is calculated using the method shown at Figure 9-10. This segment of the overall determination of NMCT first estimates the number of critical system failures that will occur during system operation over the course of a year and then calculates the time in terms of corrective maintenance that the system will not be available. This illustrates the direct relationships between reliability, maintainability, testability, and availability. As reliability and maintainability characteristics of the system design are improved, corrective maintenance requirements decrease, resulting in the system being available for operation longer.

TPM addresses the system being NMCT owing to being out of service while preventive maintenance is being performed. Calculation of TPM is shown at Figure 9-11. This formula may require expansion on an actual project to consider situations where there are several different task frequencies. In this case, each time measurement such as daily, weekly, and monthly would be calculated separately and then combined to be multiplied by M_{PMT} to determine the total TPM. This part of the overall calculation of NCMT allows for the impact of preventive maintenance to be assessed in terms of its affect on A_O. It should be pointed out that several types of preventive maintenance tasks may not be considered in this calculation. Any preventive maintenance task that does not cause the system to be NMCT

$$TCM = \left(\frac{\text{number of systems} \times \text{number of annual missions} \times \text{mission time}}{\text{MTBCF}} \right) \times M_{CMT}$$

Number of systems = the total number of systems the user can task perform a mission
Number of annual missions = the total number of times that a mission is performed
Mission time = the average time required to perform the mission once

Example: Number of systems (aircraft) = 20 (Total number in the operational inventory)
Number of annual missions = 180 (Total sorties per year aircraft)
Mission time = 8 hours
MTBCF = 200 hours
M_{CMT} = 4 hours

$$TCM = \left(\frac{20 \times 180 \times 8}{200} \right) \times 4 = 576 \text{ hours}$$

(This example is used in subsequent figures in this chapter)

FIGURE 9-10   Calculating TCM time.

TPM = number of systems × number of PM tasks × average task frequency × M_{PMT}

Number of systems = the number of systems in the inventory
Number of PM tasks = the number of preventive maintenance tasks that cause NMCT
Average task frequency = the average frequency that PM tasks are performed, i.e., daily, weekly, monthly, after each sortie, etc.
M_{PMT} = the average time required to perform a PM task.

Example: Number of systems = 20
           Average annual sorties per aircraft = 180
           Average mission time = 8 hours

PM tasks	Task frequency	Task time
Preflight	Once per sortie	0.5 hour
Postflight	Once per sortie	1.0 hour
Periodic inspection	Once per month	6.0 hours
Overhaul	1200 flying hours	40.0 hours

Calculations (for one aircraft, then the fleet of 20):

PM Task (preflight)	= 180 sorties × 0.5 hour	=	90 hours
PM Task (postflight)	= 180 sorties × 1.0 hour	=	180 hours
PM Task (periodic inspection)	=12 months × 6.0 hours	=	72 hours
PM Task (overhaul)	= (180 sorties × 8 hours)/1200 × 40 =		48 hours
Total TPM (for 1 aircraft)			390 hours
Total TPM (for 20 aircraft)			7800 hours

FIGURE 9-11   Calculating TPM time.

would be eliminated from consideration. This makes calculation of TPM more difficult that it may appear initially.

The third part of NMCT is administrative and logistics delay time (ALDT). This part of NMCT reflects the impact and responsiveness of the support infrastructure to provide adequate resources to sustain the system. ALDT consists of many factors that must be determined individually and then combined. Figure 9-12 list the factors that are commonly included in ALDT. Each of these factors plays an important role in determining the effect of ALDT on NMCT and how that effect degrades A_O.

Spares Availability

Prompt maintenance of a failed system frequently depends on spares being available when required at the location where the repair will be performed. If a necessary spare is not available, then the system

- Spares availability
- Support equipment availability
- Personnel availability
- Maintenance facility capacity
- Transportation/shipping time
- Administrative delay time

FIGURE 9-12 ALDT Factors.

will be non-mission-capable until the spare is obtained. This relationship between spares availability and operational availability illustrates how important the responsiveness of the support infrastructure is to achieving an availability goal. Procurement of spares must be linked to operational availability requirements. At the same time, it is not reasonable to assume that necessary spares always will be instantly available. The cost of the quantity of spares required to achieve this would be unreasonably high. Therefore, most procurement strategies call for a 95 percent confidence level in the spares purchased being sufficient to meet the requirements to support repair of the system. This confidence level and its contribution to operational availability will be discussed later in this chapter.

Support Equipment Availability

Many maintenance actions require support or test equipment in their performance. If the necessary equipment is not available when a maintenance action is required, then an additional amount of waiting time will be created that extends NMCT. Nonavailability of support and test equipment may be caused by failure of an item, waiting for access to an item, an item being removed to another location for calibration or repair, or insufficient quantities of support equipment that is shared by other systems. There may be instances where an item of support or test equipment is temporarily moved to an alternate location to support another system. Each of these eventualities must be considered when calculating NMCT.

Personnel Availability

Maintenance of a system requires personnel to be available to respond to random demands for performing necessary actions. These people must have the requisite knowledge and skills to perform assigned actions. Personnel availability is determined by the number of people required, the quality of the training received, and their ability to repair the system without causing further damage. These are summarized in terms of personnel waiting time and efficiency rate. Maintenance-induced error rate is a statistic that increases requirements for maintenance. A *maintenance-induced error* occurs when the person performing a maintenance action damages something else that then also must be repaired. This increases personnel requirements and lowers personnel availability.

Maintenance Facility Capacity

Maintenance actions typically are performed within a facility, such as a building or hanger. Facilities have a maximum capacity. Since maintenance facilities often are shared by many systems to make maximum use of their capacity, systems arriving for maintenance may be forced to wait until space becomes available. This is another type of waiting time that increases NMCT. Including this issue in the calculation of NMCT is often done based on the actual historical experience of the system being replaced by a new procurement.

Transportation/Shipping Time

The time required to transport the system to a location where maintenance will be performed and the time required to ship needed spares from a storage location to the system location contribute a significant portion of overall NMCT. These factors reflect the responsiveness of the transportation infrastructure to meet a specific requirement. The transportation infrastructure is shared by all systems, and therefore, its performance is measured by its overall capability rather than single events. Average shipping times normally are used when calculating this portion of NMCT.

Administrative Delay Time

The repair of non-mission-capable systems often is delayed owing to administrative requirements, such as awaiting approval to start maintenance, budget limitations, coordination requirements, and finally, simply waiting. Each of these activities may be out of the control of the owner of the system and can only be tolerated. These times are often difficult to quantify because it would require someone to track inefficiencies within the overall support infrastructure. Organizations often have standard times that are applied to calculating NMCT. When standard times are not available, then a best-guess estimate may be the only input possible. However, this additional delay in returning a system to mission-capable status is often a significant issue.

Figure 9-13 illustrates how these portions of NMCT can be estimated using appropriate calculations, standard times, and best guesses. As will be shown later, ALDT normally is a large portion of NMCT.

Figure 9-14 completes the example of calculating A_O for a fleet of aircraft. As shown by this example, everything related to supporting a system must be analyzed in the context of its overall contribution to availability. In this specific example, preventive maintenance time (TPM) is the largest contributor to NMCT. ALDT is the second largest, and corrective maintenance time (TCM) is the smallest. This gives rise to a question as to the necessity of the amount of preventive maintenance being performed. The purpose of preventive maintenance is to reduce failures. If preventive maintenance of the aircraft were reduced, there would be an increase in failures, which, in turn, would require more unscheduled corrective maintenance. This is an example of a classic tradeoff analysis where several options for supporting a system are studied to determine the most appropriate balance among all factors contributing to the overall result.

In the early stages of development of a system, A_O can be used to guide design of a system and development of its logistics support package. The way in which this is done is for the customer to provide to the contractor the anticipated values for operating and standby time, along with a historical value for administrative/logistics delay time from similar systems. The corrective maintenance time can be predicted based on the *mean time between critical failures* (MTBCF) and MTTR of the system. Note that for this calculation of availability, MTBCF is used rather than MTBF. Operational availability focuses on the probability of the system being mission-capable. Many contemporary systems are designed with redundant functions that allow them to be mission-capable after a single failure has occurred. The preventive maintenance time can be predicted based on the number of scheduled maintenance tasks to be performed over the period and the length of time required to perform each task. The combination of these factors results in a projected A_O for the system. This projected A_O is very useful in performing evaluations and tradeoff studies of design and support alternatives for the system. Typical targets for improving A_O are listed in Figure 9-15.

Fleet Availability

Operational availability, as discussed earlier, applied to a single system type or model, but it also can be used to determine requirements for a fleet of systems necessary to respond to multiple demands. This type of availability determination focuses on the size of a fleet of systems that must be mission-capable at any point in time to achieve a predetermined outcome. For example, a car rental company

$$ALDT = DT_S + DT_E + DT_P + DT_F + DT_T + ADT$$

DT_S = delay time spares
DT_E = delay time support equipment
DT_P = delay time personnel
DT_F = delay time facilities
DT_T = delay time transportation
ADT = ddministrative delay time

Example (using same example data for Figures 9-8 and 9-9)

Percent of spares available at system location	95 %
Time required to obtain spare at system location	2 hours
Time required to obtain spare from other location	72 hours
Operational availability for required support equipment	98 %
Delay time when support equipment is not available	48 hours
Average time for personnel to arrive to start maintenance	1 hour
Average time waiting to enter maintenance facility	2 hours
Average time waiting for transport to maintenance facility	0.5 hours
Average ADT while nonmission capable	4 hours
Number of maintenance actions per year (see Fig. 9-8)	144

Calculation:

$DT_S = (144 \times 0.95 \times 2) + (144 \times 0.05 \times 72)$	=	792.0 hours
$DT_E = 144 \times (1 - 0.98) \times 48$	=	230.4 hours
$DT_P = 144 \times 1$	=	144.0 hours
$DT_F = 144 \times 2$	=	288.0 hours
$DT_T = 144 \times 0.5$	=	72 hours
$ADT = 144 \times 4$	=	576 hours
Total ALDT	=	2102.4 hours

FIGURE 9-13 Calculating ALDT.

establishes a requirement to have sufficient cars available at a specific airport to respond to the needs of travelers. If sufficient cars are not available, the company will loose business to its competition. In this situation, the company must project the number of demands for cars that will occur during a specific period of time and then determine the total fleet of cars that it must have in its inventory. The fleet will include cars that are currently being rented to travelers, cars that are available to be rented, and cars that are not available. Cars not available for rental may be undergoing to routine servicing such as cleaning after return from rental, periodic servicing such as an oil change, or repairs.

$$A_O = \frac{\text{total time} - \text{NMCT}}{\text{total time}}$$

Calculating input data:

Total time = 20 aircraft × 8760 possible operating hours per year = 175,200 hours

NMCT = 576 hours + 7800 hours + 2102.4 hours   = 10,478.4 hours
               (TCM)   (TPM)   (ALDT)

$$A_O = \frac{175,200 - 10,478.4}{175,200} = 0.94 = 94\%$$

FIGURE 9-14   Operational availability example calculation.

Prediction Methods

The formulas provided earlier for the three availability statistics are for illustration of the concepts discussed. The formulas for inherent and achieved availability tend to be fairly universal in acceptance; however, calculation of operational availability can vary greatly between different organizations. The simple formula provided earlier for A_O is limited in application for several reasons. First, this formula assumes a 24/7 requirement for the system to be available for a random demand. Second, the generic formula does not include all the inputs that may be required to actually calculate the A_O for a specific system mission. Figure 9-16 illustrates the calculation of A_O for a system having a different type of mission profile. The specific formula to be used for calculating A_O for an actual system under development must be acceptable to the user. It is very important that the buyer and seller are in total agreement about the formula to be used, or the resulting calculation will be virtually useless.

- Reliability—MTBCF
- Maintainability—MTTR
- Testability—diagnostics
- Scheduled maintenance requirements
- Logistics support infrastructure
- Spares availability
- Support equipment availability
- Personnel availability
- Facility capacity and utilization rate
- Transportation responsiveness
- Administration requirements

FIGURE 9-15   Operational availability improvement targets.

$$A_O = \frac{OT + ST}{OT + ST + TCM + TPM + ALDT}$$

Where
- OT = operating time (total time is being used)
- ST = standby time (total time mission capable but not being used)
- TCM = total corrective maintenance time
- TPM = total preventive maintenance time
- $ALDT$ = administrative and logistics delay time

FIGURE 9-16 Alternate operational availability formula.

Availability is one of the primary measures of the overall goodness of a system. It incorporates many different aspects that are very important issues and presents them in terms of their contribution to the overall probability that a system will be available to perform its intended use under a predefined set of conditions when it is needed. If these predefined conditions are altered, then each aspect must be revisited to determine if the change in conditions results in a change in the operational or supportability potential of the system. After each individual aspect has been analyzed for any change, then availability is used to determine the overall effect of the change on the system and its mission capability.

CHAPTER 10
COST OF OWNERSHIP

The prediction of the total costs that will be incurred throughout the life of a system serves an important role in the acquisition process. It is a valuable aid in making decisions about different options or alternatives related to the design characteristics of the system, the support infrastructure to support the system, and the physical resources required to operate and maintain the system. The concept of *cost of ownership* is used to project the future financial obligations and liabilities that will be necessary to own the system. The use of cost of ownership during acquisition focuses on total costs over the life of the system rather than just purchase price. Supportability engineering uses various methods to predict cost of ownership during acquisition to identify significant issues that cause costs to rise so that these costs and the factors that contribute to them can be analyzed to determine ways in which they can be reduced without lowering performance or operational availability.

Cost of ownership   The total of all costs incurred to own and use a capability, including research and development costs, acquisition costs, operating costs, support costs, and disposal costs.

Three basic concepts are used by supportability engineering to estimate cost of ownership: life-cycle cost (LCC), through-life cost (TLC), and whole-life cost (WLC). Each of these methods has a different purpose and application during acquisition.

Life-cycle cost (LCC)   A technical process that compares the costs of the relative merits of two or more options.

Through-life cost (TLC)   A financial budgeting process that estimates the cost of a single option over its intended life in terms of budget category by financial accounting period.

Whole-life cost (WLC)   An estimate of the total cost to acquire, equip, sustain, and operate a single option over its intended life. WLC includes TLC plus attributable costs of the infrastructure, including the necessary training establishment and higher management and support organizations.

SYSTEM LIFE

Estimation of the cost of ownership requires an understanding of the procurement method being used and the stages through which a system evolves over its lifetime. Figure 10-1 illustrates the three generic stages in the life of a system. *Presystem acquisition* encompasses all the activities required to prepare for actual acquisition of the system. *System acquisition* covers the activities performed when the system is actually purchased and includes the cost of the system and establishing an initial support capability. *System use* includes all costs resulting from owning, using, supporting, and disposing of the system. History has shown that the largest costs of ownership occur during system use, but the majority of decisions that cause the costs to occur are made during presystem acquisition. This is why estimation of future cost of ownership during the presystem acquisition stage is extremely important to minimize the actual expenditure of money during system use. The decisions made during presystem acquisition always must consider the cost-of-ownership implications of the decision before the decision is actually made.

FIGURE 10-1   System life for purchase of an item.

A good example of system life is purchase and use of a car. In presystem acquisition, an astute consumer first decides to purchase a car as their primary method of transportation rather than taking the bus, using a taxi, or walking. This is the most significant cost-of-ownership decision because it precipitates all future costs. Then the consumer studies his or her need for a car, what it will be used to do, financial limitations, personal likes and dislikes, etc. The person also may do some market research through analysis of consumer reports and different manufacturers' information on cars available for purchase. Finally, the person will estimate the costs of different cars in terms of purchase price, operating costs, insurance, maintenance, and other significant issues. Based on the results of this presystem acquisition activity, the consumer is prepared to actually make a purchase in the system acquisition stage. After purchase, the consumer now owns the car and hopefully enjoys the results of the purchase for many years.

In some cases, the consumer or buyer may determine that no product or system that meets his or her needs is available for purchase. Therefore, a system must be specially designed. This situation is common for large items such as manufacturing facilities, military systems, mining equipment, complex or specialized systems, and other items not available on the open market. The system life then becomes more complicated because the buyer must create the requirements to which the system must be designed to meet his or her need. Figure 10-2 shows a different description of system life that is more appropriate for this situation.

The concept phase focuses on defining the detailed user need for which a system will be procured. This activity was discussed in Chapter 3.

During the assessment phase, all the reasonable options to meet the need defined in the concept phase are identified, studied, and analyzed to determine the possible outcomes should any of the

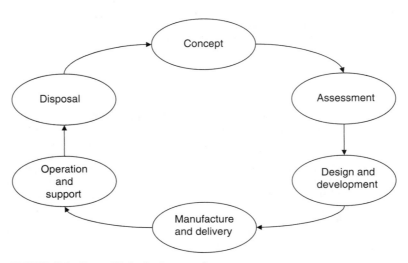

FIGURE 10-2   System life for development of a system.

options be selected. The analysis includes an estimate of operational availability and cost of owner-ship for each option. The results of the assessment phase then are tested using simulation, modeling techniques, and sometimes prototyping. Then the option that provides the best balance among perfor-mance, support, and cost of ownership is selected for procurement based on the results of appropriate evaluations and tradeoff analyses. The design and development phase incorporates all system engi-neering activities to produce an affordable design that meets the use requirements and can be supported in its intended operational environment. This phase creates the final design solution that is then passed to manufacturing.

The manufacture and delivery phase involves all detailed implementation systems engineering, design engineering, qualification testing, and manufacture of the system. It also includes identifica-tion and purchase of all the resources required to operate and support the system. At the completion of this phase, the system (or systems if more than one is produced) is delivered to the user.

The system is in actual use during the operation phase. This phase of the system life generates the major portion of cost of ownership.

The final phase of the system life is disposal. Disposal consists of removing the system from oper-ation through retirement or salvage. This normally occurs when its replacement has entered its own operation phase. Disposal also includes removal and disposal of support resources no longer required to support the system being disposed of and not required by another system or the new system.

Cost Elements

Determination of the predicted cost of ownership of a system is accomplished by combining all the relevant cost elements associated with the costs incurred throughout the system's life. Figure 10-3 shows the relationship between cost elements and the system life phases. Grouping cost elements in this manner allows the relationship between activities and the relative portion that each of these ele-ments contributes to cost of ownership. Remember, supportability engineering is responsible for influencing the design to produce a system that is as supportable as possible for the lowest cost of ownership while still having the capability to perform its intended use.

Acquisition Costs

All costs incurred from the start of the concept phase until the end of the design and manufacture phase normally are considered acquisition costs. This includes funds expended both internally and externally by the buyer, not just the funds paid to manufacturer to design and develop the system. Acquisition costs are further divided into research and development (R&D) costs and investment costs, as shown at Figure 10-3.

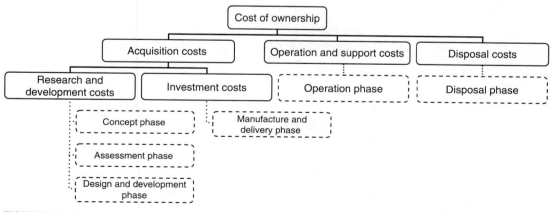

FIGURE 10-3 Cost elements.

Research and Development Costs

The costs incurred by the customer/buyer and the contractor/manufacturer during the concept, assessment, and test and select phases of the acquisition process are categorized as R&D costs. In addition, some of the early costs incurred during the engineering and manufacturing development phase are contained in R&D costs. The results of R&D is completion of the detailed documentation (e.g., engineering drawings, specifications, and plans) necessary to enter the design and manufacture phase. All costs associated with the R&D effort, including planning, management, engineering, test, evaluation, special equipment, and facilities, whether incurred by the buyer or contractors, make up this cost element.

Investment Costs

The costs of actually designing, qualification testing and producing the system, procuring and developing the necessary initial support, and establishing an initial operating capability are considered investment costs. These include the costs of building the system; initial spares and repair parts procurement; training of operation, maintenance, and supervisor personnel; support and test equipment procurement; technical documentation; maintenance facilities; and any other resources required for operation or support. Investment costs frequently overshadow other ownership costs and distort the perspective of attempting to reduce costs because they are considered the price paid by the customer to contractors to build a system. System procurement contracts can exceed a billion dollars, which seems like an enormous amount of money, but it represents only a small percentage of the total cost that someone must pay for ownership of the system over its complete life cycle.

Operation and Support Costs

The largest percentage of costs incurred over the life cycle of a system are normally due to operation and support (O&S) costs. Included in O&S costs are direct and indirect costs necessary to sustain the system. Figure 10-4 illustrates typical O&S costs elements.

Direct costs	Indirect costs
Personnel	Personnel
Consumables	Facilities
Replacement spares	Training
Support equipment	
Facilities	
Maintenance	
PHS & T	
Technical data	
Supply management	
Modifications	
Hazardous waste disposal	

FIGURE 10-4   O&S costs.

Direct O&S Costs.   Any cost that has a direct relationship to the operation or support of a system is considered a direct cost. Costs for personnel include operators, maintenance personnel, and supervisors who are responsible for the operation or maintenance of the system or components. These can include support personnel who are directly related to the mission scenario of the system. A subelement of personnel cost is the cost of specific training related to operation or maintenance of the system, which can be either initial training or sustainment training. Consumables are any items that are required to sustain operation or maintenance (e.g., fuel, lubricants, maintenance materials, expendable supplies, or repair parts). The purchase of spare parts needed to replace initial provisioned items or increase the range and depth of spares is a direct cost. The cost of maintenance of support and test equipment and procurement of replacement items is also a direct cost. Any facility costs that are incurred for support of operation or maintenance activities are direct, but this does not include the construction or modification costs, which are considered an investment cost. Facility costs also include the cost of water, power, and other utilities directly related to maintenance or operations. Maintenance of supplies and equipment is a direct cost, but the labor requirements for maintenance are included in personnel costs and should not be duplicated in this cost element. Packaging, handling, storage, and transportation (PHS&T) costs include all movements of the system owing to operation or maintenance needs after initial delivery and the movement of spares and repair parts between maintenance facilities, supply facilities, and the user. Technical data initially are procured as an investment cost, but maintenance and updating of the data are a direct O&S cost. Supply management costs are attributed to the unique spares and repair parts of the system that must be stocked to support operations or maintenance. These costs are incurred at all levels of supply from organizational to depot and also include the administrative costs of maintaining records for the items. All engineering changes and other modifications to the system that occur after deployment are direct O&S costs. Modification costs are considered sustaining investment costs that are necessary to enhance the reliability, maintainability, supportability, or operational capabilities of the system.

Indirect O&S Costs.   Costs that are incurred for relevant services, support personnel, and noninvestment items that are necessary to sustain operations or maintenance but cannot be related directly to a specific system are categorized as indirect O&S costs. These costs can include a broad range of cost elements such as military installation facilities, medical facilities, maintenance of real estate, and initial training costs. Personnel costs classified as indirect may include medical personnel, initial training instructors, and personnel administration and management. Indirect facility costs consist of real property maintenance and upkeep, installation maintenance, base exchanges, and commissaries, and other facilities that are required to indirectly support either the personnel or operation and maintenance of the system. Initial military training cannot be attributed to a specific system but is required to produce trained operation and maintenance personnel.

Disposal

A cost element that is often ignored is the cost of disposing of a system as it becomes obsolete or is replaced. In some instances, the equipment may have salvage or resale value that may off set the cost of disposal, but costs can be incurred. Figure 10-5 lists typical costs that occur during disposal. Spares and repair parts that are unique to the system being disposed of must be purged from the active supply inventory, which may constitute a significant expense. If the system, such as an aircraft, has many spares and repair parts that must be disposed of, then the cost for such an operation should be identified. PHS&T costs are incurred to physically move the system and support resources from operational sites to a disposal site. As a part of the disposal effort, the data collected during the life cycle of the system must be closed and dispositioned. Significant data related to operations, reliability, maintainability, performance, or other information that has other uses are reviewed and forwarded to the appropriate destination. If the system is to be resold or redistributed to other users, then refurbishment or overhaul may be required. A portion of this cost may be recouped after transfer, but portions may be charged as a part of disposal.

- System disposal or refurbishment
- Recycling
- Inventory closeout
- Data management and archiving
- Disposal of hazardous materials

FIGURE 10-5 Disposal costs.

Demilitarization is the act of rendering an item useless for military purposes. Government regulations require that certain classes of items be demilitarized before disposal. If the system being disposed of requires such actions, then the costs are accrued as disposal costs. Systems or their components that contain dangerous or hazardous materials require special handling for disposal. If the item contains nuclear materials or dangerous chemicals, then the disposal process may be lengthy and very costly. Such costs may have a significant impact on the predicted life-cycle cost.

The total cost of ownership that is estimated using the preceding cost elements may produce a surprising result. The actual cost of ownership of every system is different and may vary greatly, but the ratio between R&D, investment, O&S, and disposal costs for most system tends to be similar. This similarity has been the subject of many studies. The general consensus of these studies is presented in Figure 10-6. This figure suggests that for an average system, 2 percent of cost of ownership occurs during R&D, 12 percent during investment, 85 percent during O&S, and 1 percent during disposal. These studies also suggest an even more important point: When decisions are made affects the cost

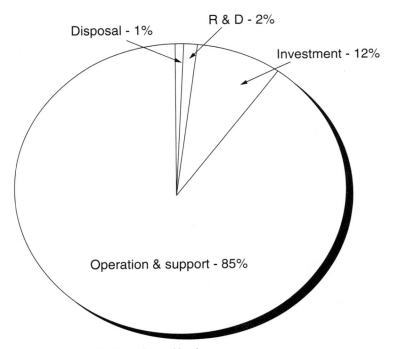

FIGURE 10-6 Ratio of costs incurred by phase.

Phase	% of decisions effecting cost of ownership
Research and development	70%
Investment	20%
Operation and support	10%
Disposal	0%

FIGURE 10-7   Ratio of decisions affecting cost by phase.

of ownership. Figure 10-7 shows the percentage of cost-of-ownership decisions made in each phase that dictate future expenditures of money.

ESTIMATING COST OF OWNERSHIP

At the beginning of this chapter, three different concepts for estimating cost of ownership were introduced: life-cycle cost (LCC), through-life cost (TLC), and whole-life cost (WLC). A review of the definitions provided shows that each technique has a different purpose. On a development program, LCC is used before the design solution is known, and TLC is used after the design has been completed. A model for estimating TLC requires detailed data about each item within the architecture of the system. This is only available after the system is designed. LCC normally only addresses issues at the system level in order to ensure that the right design characteristics are included in the specification to which the system will be produced. Before attempting to estimate cost of ownership, a decision must be made as to which concept will be most useful to aid in the decision-making process.

Life-Cycle Cost

LCC is a comparison of different options or alternatives. The purpose of LCC is to highlight these differences. In order to do this, any LCC estimate should focus on factors or characteristics of each option that have different effects on cost. Any factor that is identical between the options may be eliminated from the estimate to aid in highlighting differences. The resulting estimated costs then can be used to determine the most cost-effective option. This means that conceptually, LCC does not actually attempt to produce an exhaustive and totally complete cost estimate. For example, when a person is deciding which car to buy, he or she will make an estimate of the cost of each option based on the differences between each car, such as fuel consumption rate, insurance, maintenance, and servicing. Costs that would be the same regardless of which car is purchased, such as parking, tolls, and washing, would not be included in the estimate because they would be the same for either. By eliminating the factors that have no participation in highlighting the differences, an LCC model is brief, focused, and requires a minimum quantity of input data to produce an estimate. LCC produces a bottom-line number that allows the difference to be compared. It does not attempt to estimate the total cost of ownership.

This concept was discussed extensively in previous chapters. In the R&D phase of acquisition, very little detailed, accurate data about the final design solution is available. Therefore, a model that estimates LCC must be capable of accepting estimates, rough data, and guesses. This allows global optimization and conducting "what if" scenarios quickly to see different results. An LCC model may be capable of producing a reasonable result with as few as 100 data inputs. Most of these data inputs come from the use definition, existing similar systems, and best engineering judgment. LCC is the primary tool used by supportability engineering to assess the impact of design characteristics while participating in the system engineering process. It is most beneficial when used during windows D1, D2, D3, and D4 (see Figure 2-8) on design programs and windows O1, O2, and O5 (see Figure 2-15) on off-the-shelf (OTS) programs. This focuses decisions about the relative costs that might be realized for each option under consideration. LCC produces relative costs, not actual final expenditures of money. LCC is not about exactness of the answer produced. LCC is designed to identify the most significant factors that will contribute to actual ownership of an item.

Through-Life Cost

TLC is a budget-estimation technique that attempts to project the actual expenditure of money over the life of a system. The results of this technique are used to establish the actual budget for the system. Therefore, it should be as complete and detailed as possible. TLC produces an output that projects the budget for each category of expenditure by financial accounting period. A model that estimates TLC is extremely detailed and requires a vast amount of very accurate input data to produce a usable output. The definition provided earlier stated that TLC estimates the cost for a single option. Using the same example of a person buying a car, he or she might perform a TLC estimate to determine the portion of his or her personal budget that would be required each month to operate and support a particular option. In this case, the cost of parking, tolls, and washing would be included because the person would have to have sufficient money available to pay each expense in the month when it occurred. TLC modeling can be done only after the design of an item has been determined. TLC is most beneficial during windows D5 and D6 (see Figure 2-8) on a design program and during window O5 (see Figure 2-15) on an OTS program.

Whole-Life Cost

The concept of WLC is relatively new. It is an attempt to estimate the overall cost of ownership for a system by combining TLC with other costs that are above the system being procured. These other cost include anything that must be in place in order for the system to be effective. This is an exhaustive list of anything that is necessary to provide a management, administrative, and support infrastructure. WLC can be confusing. Both LCC and TLC address the direct and indirect costs of a system. They do not address costs outside the project that they have no control over or have no way of changing. There are many outside costs that are shared over many individual projects. The military provides a good example to illustrate WLC implications. Maintenance personnel are required to operate and maintain a system. Direct and indirect personnel costs are addressed by LCC and TLC. However, the military incurs additional costs for personnel. These include cost of basic training on how to be a soldier, the cost of medical care, the cost of accommodation, the cost of meals, the cost of transportation, the cost of relocation, etc.

Determining the personnel segment of WLC necessitates identification of every aspect associated with having the necessary people present. Typically, WLC is used occasionally by supportability engineering as an overall projection of costs for a large development program. It is a high-level management tool that is used most effectively to see the total cost of a capability rather than the cost of ownership for a system. One of the most beneficial applications of WLC is to determine affordability of any system to meet a defined need. WLC may be used during window D1 (see Figure 2-8) on a design program or during window O1 (see Figure 2-15) on an OTS program to assess affordability or to establish a budget for a new program based on the actual costs for a previous system or the system being replaced.

COST-ESTIMATION MODELS

Cost-estimation models are by nature extremely complex if they are to be of any useful value. As described earlier, there are a myriad of cost factors that must be considered when attempting to predict the cost of owning a system. There are many models available for use; however, an organization should have one or two preferred models that should be used consistently on all projects. In each case, the models should have been developed to address situations related to operation and support of the specific product or system types of the user. Proper use of cost-estimation models depends on an understanding of the intent of the model, how the model deals with input data, how the model actually calculates costs, and the philosophical views of the model builder by whom the model was designed. Never use a cost-estimation model unless you know the algorithms. Some models are actually a series of submodels that address certain aspects of the life cycle.

Modeling Concept

The basic concept of cost modeling is illustrated at Figure 10-8. Each of the major cost elements can be expanded to include several hundred subelements and variables. Figure 10-9 shows how the concept of Figure 10-8 can be expanded using only the subelements identified in the preceding section of this chapter. This refinement process can be repeated until the resulting model contains elements that address every cost that can be associated with a system.

Modeling Problems

When cost-estimation models are used as a tool for making critical design and support decisions, problems can occur that distort model utility. These problems should be considered when choosing the model and interpreting the resulting predictions. Common problems include use of invalid assumptions when insufficient data exist, changes in production schedule or order quantities, lack of uniformity in categorizing cost elements, inadequate description of the life cycle, use of obsolete data, inappropriate cost element structure, and use inaccurate inflation or discount rates. Any combination of these problems can invalidate the results of a model. Another common problem associated with using the results of models is for analysts to focus too much attention on the cost aspects rather than the limited availability of some critical resources. Sometimes cost may not be the driving factor for making critical support decisions; it may be the optimal use of limited critical resources. If an LCC model is to be a useful tool in analyzing the total cost of a system, it should contain certain characteristics. Regardless of the type or origin of the model chosen, it should be capable of providing comparisons and evaluations for tradeoff analyses of alternative options, identification of risks, and establishing a baseline for sensitivity analyses throughout the acquisition process.

Estimating Techniques

There are three accepted techniques for estimating costs when data for modeling are not available. These techniques are analogy, parametric, and engineering estimating. Each method has varying degrees of application to the phases of a system life cycle. *Analogy-based estimates* normally are

$$C_T = C_R + C_I + C_O + C_D$$

C_T = total cost of ownership
C_R = research and development cost
C_I = investment costs
C_O = operation and support costs
C_D = disposal costs

FIGURE 10-8   Cost model.

$$C_T = C_{RP} + C_{RM} + C_{REN} + C_{REV} + C_{REQ} + C_{RF} + C_{IPR} + C_{IPN} + C_{IM} + C_{IS} + C_{ISE} + C_{IM}$$
$$+ C_{IE} + C_{IF} + C_{IP} + C_{ODP} + C_{OC} + C_{ORS} + C_{OSE} + C_{ODF} + C_{ODM} + C_{OP} + C_{OTD}$$
$$+ C_{OSM} + C_{OM} + C_{OIP} + C_{OIF} + C_{OIT} + C_{DI} + C_{DP} + C_{DDM} + C_{DR} + C_{DD} + C_{DW}$$

C_T	=	total life cycle cost
C_{RP}	=	R&D planning costs
C_{RM}	=	R&D management costs
C_{REN}	=	R&D engineering costs
C_{REV}	=	R&D evaluation costs
C_{REQ}	=	R&D equipment costs
C_{RF}	=	R&D facilities costs
C_{IPR}	=	Investment production costs
C_{IPN}	=	Investment planning costs
C_{IM}	=	Investment management costs
C_{IS}	=	Initial spares cost
C_{ISE}	=	Initial support equip. costs
C_{IM}	=	Technical manual costs
C_{IE}	=	Investment engineering costs
C_{IF}	=	Investment facilities costs
C_{IP}	=	Initial PHS&T costs
C_{ODP}	=	O&S direct personnel costs
C_{OC}	=	O&S consumables cost
C_{ORS}	=	O&S replacement spares costs
C_{OSE}	=	O&S support equipment costs
C_{ODF}	=	O&S direct facilities costs
C_{ODM}	=	O&S maintenance costs
C_{OP}	=	O&S PHS&T costs
C_{OTD}	=	O&S technical data costs
C_{OSM}	=	O&S supply management costs
C_{OM}	=	O&S modification costs
C_{OIP}	=	O&S indirect personnel costs
C_{OIF}	=	O&S indirect facilities costs
C_{OIT}	=	O&S indirect training costs
C_{DI}	=	Disposal inventory closeout costs
C_{DP}	=	Disposal PHS&T
C_{DDM}	=	Disposal data management costs
C_{DR}	=	Disposal refurbishment costs
C_{DD}	=	Disposal demilitarization costs
C_{DW}	=	Disposal waste management costs

FIGURE 10-9 Cost model elements.

used very early in the concept phase when detailed information on a new system is not available. Analogy-based estimating uses historical information on similar systems as a basis for developing estimates for the new system. These estimates tend to be top level at system and subsystem levels. *Parametric estimating* may be developed in the early design phases of a program where commonality of the new design can be gauged based on an existing design where material and labor content for manufacture of the old system are known. A parametric estimate allows more details of the lower levels of a system to be addressed. *Engineering estimates* are used when detailed physical characteristics of the new system are known, but no field experience is yet available. These techniques can be used independently or collectively to produce estimates of the predicted cost of ownership.

CREATING A COST-ESTIMATION MODEL

There is no set form or context for development of a cost-estimation model. It is not a scientific process but rather an art. This means that to be successful, a cost-estimation model must fit the situation being studied. Philosophically, all cost-estimation models are the same, but in actual fact, every one is different. This means that the exact same input data can be put into two models, and the results will be wildly different. Thus extreme caution should be taken anytime cost estimation is being performed. Selection of the appropriate model is crucial. There are many brand-name cost-estimation models available for purchase, and they all will create a final cost estimate, but each will be philosophically different. It may be more advisable for an organization to develop its own model to fit its specific requirements. Either way, the user of a cost-estimation model must have access to the formulas behind the model and understand how the model models things.

Cost Types

There are two cost types that are included in cost-estimation models: nonrecurring and recurring. A *nonrecurring cost* is one that happens only once. Including this type of cost in a model is fairly simple because all that is required is to identify what the cost is and when it happens. Examples of nonrecurring costs are provided at Figure 10-10. Complexity of cost-estimation models is due to recurring costs. *Recurring costs* are much more difficult to model because they repeat with some frequency, and the model must determine what the cost is and when it occurs because of this repetition. Figure 10-11 presents examples of recurring costs. The formulas required to calculate recurring costs normally require several input values of different types from different sources. In the example provided at Figure 10-12, the annual cost of fuel for a car is calculated by dividing the annual mileage by the engine fuel consumption rate, which then is multiplied by the price of fuel. Each of these input

- Design of system
- Manufacturing setup
- System purchase
- Initial spares
- Facility construction

FIGURE 10-10   Nonrecurring costs.

- Fuel costs per year
- Maintenance labor costs per year
- Materials costs per year
- Servicing costs per year
- Operating costs per year

FIGURE 10-11   Recurring costs.

values is different. The annual mileage is determined by requirements to drive the car. The fuel consumption rate is related directly to the design of the engine. And the price of fuel is totally independent. The annual mileage is determined by the user. The fuel consumption rate is determined by the car designer. The price of fuel is due to supply and demand in the marketplace. The user has direct control over the number of miles driven and can vary this factor over the life of the car. The user probably selected the car but had only one opportunity to decide. The user has no control over the price of fuel. The annual cost of maintenance may also be a significant issue when deciding which car to buy. There are many factors used to predict these costs, but the majority of the input values relate to the planned use of the car. The basic concept of this idea, shown in Figure 10-13, is that the more a car is used, the more it will require maintenance and so the resulting maintenance costs can be estimated based on the planned use of the car. Recurring costs calculations take many different values and combine them in such a manner as to produce a representative cost prediction.

Input Data Types

There are two basic types of input data that are required for cost-estimation modeling: constants and variables. A *constant* input is a value that does not change over the life of a system. An example of a

Formula:

Annual fuel costs = (miles driven per year/miles per gallon) × fuel cost per gallon

Input data:

Miles driven per year	- 12,000
Miles per gallon	- 20
Fuel cost per gallon	- $3.00

Calculations:

Annual fuel costs = (12,000/20) × 3.00 = $1800.00

FIGURE 10-12   Calculating annual fuel costs.

Formula:

Annual maintenance labor costs = (miles driven per year/MTBF) × MTTR × Labor cost per hour

Input data:

Miles driven per year	- 12,000
Mean time between failures	- 500 miles
Mean time to repair	- 4 hours
Maintenance labor cost	- $50 per hour

Calculation:

Annual maintenance labor costs = (12000/500) × 4 × 50 = $4,800.00

FIGURE 10-13   Calculating annual maintenance labor costs.

constant would be the number of systems owned. A *variable* input is one that changes over time owing to some influencing factor. An example of a variable input would be the inflation rate. Examples of both constant and variable inputs are provided at Figure 10-14. Every input for a cost-estimation model will be either constant or variable. Care must be taken when expressing values to ensure that the proper measurement base is stated. In the example in Figure 10-12, if the fuel consumption rate was stated in miles per gallon and fuel cost was stated in cost per liter, then the resulting calculation would be in error unless the fuel consumption rate was translated into liters or the fuel cost was converted into gallons. It is a common error to ignore the consistency of input data, which later causes false results from the model.

Constants	Number of systems
	Number of operators per system
	Operating locations
	Maintenance locations
	Length of training course
Variables	Inflation rate
	Annual operating time
	Mission duration
	Maintenance times
	Usage rate
	Cost of materials

FIGURE 10-14   Input data type examples.

Model Coverage

The developer of a model must determine what areas of cost of ownership the model actually will cover. Most models focus on operating costs and costs of supporting hardware corrective maintenance. These areas tend to be fairly easy to incorporate into any modeling process. The areas that are more difficult are software maintenance, preventive/scheduled maintenance, and disposal. Software maintenance is difficult because software does not wear out or break like hardware; therefore, it is difficult to use the hardware support modeling methods on software. The only way that has proven to work for this situation is to estimate the number of faults or errors that must be corrected and then add the additional effort for increasing functionality. Preventive or scheduled maintenance is performed based on either system usage, calendar time, or other events. These increments of use vary from determining the number of failures to be repaired based on mission and reliability. Finally, disposal is far into the future, and very little is known about disposal problems at that future point. Thus most models only estimate the cost of disposal as a percentage of system cost in terms of salvage. These three areas can be very significant in the cost of owning a system, so it is important that a cost-estimation model adequately address each area when applicable. Model developers must establish their philosophical view on each area before starting to produce formulas for the model. It is very important that any model user determine how the model he or she is about to use addresses each area, or the model results may be useless for their decision making.

Using LCC and TLC Models

As discussed earlier, this is an inexact process that attempts to gather and use estimates, assumptions, and historical information to predict what may happen in the future and translate the results into cost. Therefore, why do LCC? The answer is that LCC, although imperfect, is the only tool available to supportability engineering to assess the impact of design, operation, and support decisions on the total program. This is the classic "what if" tool. LCC modeling allows supportability engineering to create a reasonable projection of cost based on an assembled data set and then to look at the impact of changes on the baseline data. What if this or that factor changes? What is the effect of the change on other factors? What happens to the overall projected LCC? This is life-cycle costing. Supportability engineering actually may use several different LCC models over the course of an acquisition program. In the early concept phase, LCC models tend to be simple and capable of accepting minimum input values to calculate global estimates, sort of rough order of magnitude. These types of LCC models are those used most often to influence the early design of the system by determining the correct design criteria to be included in the product specification.

When the program moves into detailed design, a TLC model tends to be much more complex and require large amounts of input data. These detailed TLC models are for resource optimization rather than design influencing. Choosing the correct model for the specific application being modeled is crucial to having any useful results. LCC modeling is the tool used most often by supportability engineering when making tradeoff analyses while participating in the systems engineering process.

Operational Availability, Life Cycle Cost, and Operational Effectiveness

Decision making during the acquisition of a system must be based on the best possible information. However, the most critical information is about what might happen in the future, and this is unknown. Modeling techniques provide the ability to project the future. The use of operational availability and LCC provides the ability for supportability engineering to assess alternatives and options for their relative merits. This is critical for ensuring that the final design solution meets its required performance objectives, that it will be available to perform its intended use when required, and that it will be cost-effective. By combining operational availability and LCC into an overall projection, supportability engineering is able to achieve a reasonable balance among the quality of the system, the support infrastructure for the system, and the cost of ownership.

Options being considered	O_E	A_O	C_O
Diesel engine	94	89	$100
Gasoline engine	91	92	$90
LPG engine	88	94	$80

Where O_E is rating of capability out of 100 points
A_O is probability of being mission-capable
C_O is cost of ownership per day

FIGURE 10-15   Decision-making criteria.

Figure 10-15 illustrates how these statistics are used in the decision-making process. The tradeoff being illustrated is a decision as to which type of engine will be used in a vehicle. If operational effectiveness is the only decision criterion, then the diesel engine is the obvious choice. If cost of ownership is the only decision criterion, then an LPG engine is the choice. If operational availability is the only decision criterion, then again, the LPG engine is the choice. But if the decision is to achieve the most reasonable balance among the three factors, then the gasoline engine is probably the choice. This is the ultimate use of modeling.

CHAPTER 11
SUPPORTABILITY ANALYSIS

Supportability   A *prediction* or *measure* of the characteristics of an item that facilitate the ability to support and sustain its mission capability within a predefined environment and usage profile.

The supportability characteristics of a system are expressed in terms of goals, objectives, thresholds, and constraints. Ultimately, these must be verifiable in the final design solution. Supportability analysis provides the methodology. Six types of activities are performed by supportability engineering, and they are listed in Figure 11-1. These activities must be performed in the order listed to realize the total benefit of supportability analysis.

Supportability engineering performs various interrelated analyses throughout the life of a system. These analyses commence as an integral part of the systems engineering process during the earliest activities of system acquisition. They must be a continuing focus throughout the acquisition process. As the design matures, the focus shifts to determining the physical support requirements. Finally, supportability engineering uses in-service experience to identify requirements and opportunities for design upgrades and improvements that will enhance supportability.

Supportability engineering performs a series of analyses that combine system design characteristics, system use, support infrastructure, and support resources into a single optimized solution that achieves system performance requirements while at the same time minimizing total cost of ownership. The activities listed in Figure 11-1 encompass all these aspects. First, there must be clear and unambiguous definition of supportability requirements for the system. These requirements then are translated into supportability design characteristics. After the design characteristics are clearly identified, they are expressed in measurable goals, thresholds, and constraints. A *goal* is defined as an aim or desired outcome. A *threshold* is a minimum that must be attained. And a *constraint* is something that must be done or cannot be done. Supportability engineering then assists system engineering and design engineering in implementation of these measurable characteristics in the final design solution. As the design moves into its final stages, a different type of supportability analysis is performed to identify support resource requirements, and then the quantities of each resource that must be provided to support the system are determined.

FUNCTIONAL THEN PHYSICAL

The system design process starts with a functional set of requirements—what the system must do—and ends as a physical entity that will perform all necessary functions. This divides all activities for supportability analysis into two very different phases. Analyses are performed as a part of system engineering to define and implement the functional capabilities of the system, and then a different type of analysis focuses on supporting the design solution. Most people who have experience with supportability analysis are familiar with the latter, dealing with supporting the system, rather than the former, functional requirements identification and implementation. However, 70 percent of decisions affecting cost of ownership are made during the functional stage of system development, as has been

- Definition of supportability requirements
- Formulation of supportability design characteristics
- Establish supportability goals, thresholds, and constraints
- Assist in implementation of design characteristics
- Identification of support resource requirements
- Determination of support resource quantities

FIGURE 11-1   Supportability analysis.

stated in Chapter 10. Therefore, it is imperative that supportability analyses are an integral part of this decision-making process to ensure that the final design solution achieves a balance among performance, supportability, and cost of ownership.

FUNCTIONAL SUPPORTABILITY ANALYSIS

The application of supportability to the functional requirements of the system should create a progressive sequence of events that starts with the user's need and ends with a system design solution that possesses all the requisite characteristics that will allow the system to be supported. Figure 11-2 shows this sequence of events.

User Need–System Requirements Study

The first event is determining the specific issues of the user's need that form necessary input into supportability analysis. The system requirements study, discussed in Chapter 3 and illustrated at Appendix A, is the vehicle for this determination. Supportability engineering provides a significant portion of the inputs to this study. This is not an analysis; rather, it is gathering existing information about the user's need and presenting the information in an organized manner that provides a single starting point for supportability analysis. The first activity of supportability engineering on any acquisition project should be to participate in preparation of these portions of the system requirements study. Remember, the study provides the starting point for analysis, not the results of the individual analyses. The study is simply a collection of existing information and assumptions.

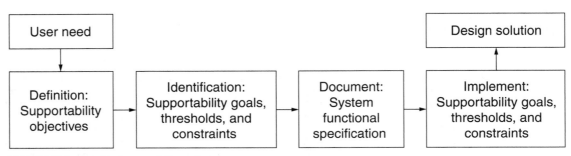

FIGURE 11-2   Functional supportability analysis.

Definition of Supportability Requirements

At the start of an acquisition project, the new system does not exist; only its function requirements exist. The system engineering process described in Chapter 4 illustrated how a functional description of the system is developed into a functional block diagram. This diagram addresses performance functions. Supportability engineering participates in this development and then adds to the diagram potential support functions. Initially, support functions are identified based on the technology baseline of the new system and then are refined through comparisons with existing similar systems. For example, a car with a gasoline engine will require a specific set of support functions based on the technology of the engine. A car that is powered by an electric motor will have different support requirements. A car with a steam engine will have requirements for support that are different from the gasoline engine and the electric motor. This is the starting point for identification of the possible support requirements for the new system based solely on its technology baseline. The next step in the definition of supportability requirements is comparison with existing, similar systems.

Comparison System

Supportability engineering takes the performance functional block diagram and searches for an existing system that has the same or very similar performance functions. The rules for selecting a comparable system for analysis are listed at Figure 11-3. The first system considered for selection is the system being replaced. It is the most appropriate and should be used unless its technology baseline is obsolete and does not provide a reasonable comparison of future requirements for the new system. If the system being replaced is determined to be unacceptable for comparison, then supportability engineering must identify a similar system that has a comparable technology baseline.

Comparison Analysis

A comparison analysis is performed on the selected comparable system. There are two aspects to this analysis: defining support requirements for the comparison system and identification of significant supportability and support issues. The definition of support requirements starts with determining how the comparison system is currently maintained. Figure 11-4 shows an example of a support philosophy. This diagram illustrates the overall approach used to support an item when it fails and the flow of materials between each location to support the process. This is a high-level depiction of how the system is supported, but it does not show the actual functions required to achieve this philosophy. The next step is to determine the support functions performed at each location. The support requirements for a system are generated as the system is operated and the need to maintain the system in an operable condition. Figure 11-5 shows an example of the support functions

1. Comparison system performance functions must be similar to new system
2. Comparison system must be currently in operation
3. Comparison system operational environment must be very close to operational environment of the new system
4. Comparison system must have sufficient use history to provide definitive support requirements
5. Comparison system support environment must be similar to existing or adaptable to new system

FIGURE 11-3   Comparison system selection rules.

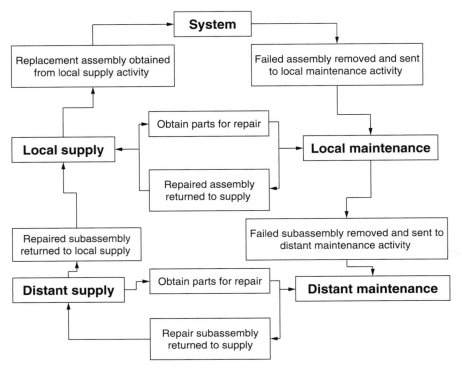

FIGURE 11-4 Support philosophy.

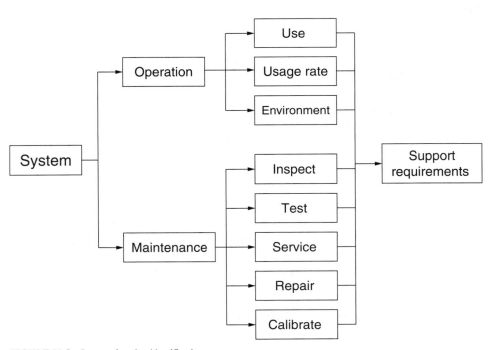

FIGURE 11-5 Support function identification

required to support a system with a given technology baseline. The combination of all operation and maintenance activities drives requirements for support resources. If the new system and the comparison system have the same technology baselines and their uses and environments are similar, then the functions required to support the new system will be very close to those of the comparison system. This first step of comparison analysis determines the functions that will be required to support the new system.

The second step of comparison analysis is to learn about the successes and problems of the comparison system through application of the modeling techniques discussed previously in this text. Figure 11-6 shows the process for performing a comparison analysis using operational availability and life-cycle cost to develop supportability requirements for the new system and identification of any risks or limitations on providing the minimum support needed for the new system.

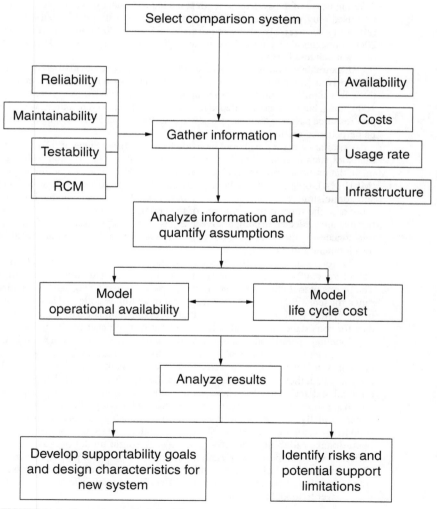

FIGURE 11-6   Comparison analysis modeling process.

A comparison analysis, by definition, compares the differences and similarities between two different options. A supportability comparison analysis compares the differences and similarities between what the comparison system was required to achieve when it was developed and what it actually experienced when used. Each piece of information gathered for the comparison system receives this type of analysis.

An easy understanding of this technique can be provided using the mean time between failures (MTBF) of the comparison system. For example, a comparison system was required to achieve an MTBF of 1000 hours. However, in actual use, it only achieved 700 hours. This is an interesting observation, but the numbers themselves do not provide any answers, only more questions. The first question should be, "Was the system used within the constraints for which it was designed?" If system use exceeded its intended use profile or environment of use, then the difference between required and actual MTBF may have been due to this. It is a problem for further investigation. The owner of the comparison system probably used more support resources than were anticipated and thinks that a larger quantity of resources is necessary. This is true, but the cause for an increase in resources is the different usage of the system rather than inadequate quantities of resources. If the comparison system was used only within the conditions for which it was designed, then it has a design problem with reliability. Conversely, if the required MTBF was 1000 hours and the system actually achieved 2000 hours, this also requires investigation. First, was the system used to the full extent it was designed or to a much less harsh profile?

An example would be a truck being designed for extreme off-road driving and then only actually being driven on the highway. If this is true, then the system may have been overdesigned. However, there is a significant problem: The support infrastructure for the system was based on an MTBF of 1000 hours, but the system failed at half that rate. Money and resources were wasted in creating a support resource package that was grossly overstated and underused. This is a real problem that must not be perpetuated on the new system.

A similar comparison should be made for every basic piece of information used by this analysis. Statistics such as mean time to repair (MTTR), fault detection and isolation rate, false alarm rate, mean labor hours per operating hour (MLH/OH), mean labor hours per maintenance action (MLH/MA), operational availability, average cost of spares, average cost per maintenance action, and administrative and logistics delay time (ALDT) provide valuable insight into the successes and failures of the support infrastructure of the comparison system. Figure 11-7 is a list of typical statistics that are subjected to comparative analysis. This analysis assists in identification of significant factors concerning the design and use of the comparison system that may be targets for improvement in development of the new system.

The total range of statistics needed to perform a comparative analysis may not be available. Therefore, reasonable assumptions must be made to provide input information that is not available. The best way to do this is by determining a reasonable range of possible statistics and then using the range for sensitivity analysis once the initial modeling is done. All assumptions must be documented in the system requirements study so that all organizations performing analyses have the same base data for consistency across all outcomes of various engineering disciplines.

Modeling operational availability and cost of ownership, as described in Chapters 9 and 10, provides the method of consolidating all the individual statistics into an overall view of what the comparison system was suppose to achieve with what it actually achieved. After the initial model results are generated, they are analyzed to identify the significant issues that caused the difference between required and actual results. These results should be subjected to a sensitivity analysis to determine which are significant. Some will be very significant, having a major contribution to the system results. Others actually will be insignificant in terms of causing the success or failure of the system and its support infrastructure. It is desirable that the model used for the comparison analysis incorporate both operational availability and life-cycle cost. A single model provides a consistent philosophical definition of the importance of each input statistic and how it participates in the overall model result. The importance of selecting the appropriate model cannot be overemphasized. In most cases, a spreadsheet-based model created or modified for each specific acquisition project proves to be the best answer to model selection.

Systems	Number of systems supported
	Number of operating locations
	Number of maintenance and support locations
System usage	Mission time
	Operating time
Cost	Unit price of system
	Average spare cost per repair
	STE cost
	Personnel per repair costs
Reliability	MTBF
	MTBMA
Maintainability	MTTR - 1st level
	MTTR - 2nd level
Testability	Automatic fault detection & isolation percent
	Manual fault detection & isolation percent
	Automatic test equipment requirements
Availability	A_I
	A_A
	A_O
Supportability	Administrative and logistics delay time
	Spares confidence level
	Maintenance proficiency level

FIGURE 11-7   Comparison statistics.

Comparison Analysis Results

The purpose of comparison analysis is to identify the significant issues that must be considered in the development of a new system. Comparative analysis is not designed to solve all the problems of the new system. Its purpose is to highlight the things that have made the biggest contribution to success or failure, sometimes called the *drivers*. Comparison analysis is always done at system level, the level where a specification is written. It is rarely done for individual assemblies within a system because the physical design of the new system does not exist at this stage of development. Results of comparison analysis should be stated in three categories: (1) design characteristics that affect supportability,

(2) support infrastructure adequacy and limitations, and (3) support resource requirements and utilization. These become the focus of supportability engineering for the new system.

Standardization Analysis

The purpose of a standardization analysis is to identify potential options for development of a system design that minimizes requirements for support resources. There are two sources for information used in this analysis: (1) resources that will be available to support the new system when it is placed into service, as identified in section 4 of the system requirements study, and (2) resources currently used by the system being replaced that will become available when no longer required to support that system. *Standardization* simply means limiting the range of items required to support the system. Limiting the range of items normally reduces the size of the support infrastructure necessary to support the system. This analysis must be performed before the start of actual design engineering activities. The results of this analysis form an input to the design instructions written by system engineering that are followed by design engineers.

Standardization analysis investigates each type of support resource to identify opportunities for its application to the new system design. Figure 11-8 lists typical support-resource categories that are analyzed. These categories include hardware, software, and support resources. Supportability

Hardware:

- Assemblies
- Materials
- Fit
- Access
- Left/right interchange

Software:

- Language
- Modules
- Conventions
- Note coding

Resources:

- Parts
- Tools
- Support equipment
- Test equipment
- Facilities
- Personnel
- Training
- Packaging
- Transportation

FIGURE 11-8   Standardization analysis categories.

engineering does not perform this analysis to force a specific design attribute; it does this analysis to identify options for design engineers to pursue that eventually will result in a more supportable design.

Hardware Standardization.   Various hardware options are investigates to identify possible standardization opportunities. This may include identification of a common power supply or pump that can have multiple applications rather than having several different items that perform the same function. A standard grade of materials used throughout the system may lower production costs. It is desirable to have a common fit of items, such as having all circuit cards fit the same in the system so that any circuit card can be removed and replaced using a single repair procedure. The system should be designed with accessibility to mission-critical items. Chapter 5 discussed using individual item failure rates to position the least reliability items in front of those that will not fail as frequently. Standard access covers and access panels tend to make maintenance procedures shorter. Where a system has like items installed on different sides, such as engines on an aircraft, the right installation and the left installation should be identical. This provides for commonality in inspections and maintenance procedures and for common support resources.

Software Standardization.   Standardization issues for software design are extremely important for through-life support. A single language should be used for all application software in the system. This allows consolidated planning for its support. Common software modules should be used and reused throughout the system, where possible. A single-code functional flow should be maintained rigorously for all software modules. A standard coding convention should be selected and adhered to when coding software instructions. Finally, a standard method of note coding (comments inserted in the code describing each function) should be followed for all system software. These standardizations features will allow the system software to be more cost-effective to support for the life of the system.

Resource Standardization.   A standardization analysis always should be performed for resources on a design program. The purpose of this application is to limit the range of resources required to support the system when in operation. The design engineer creates requirements for resources when a system design is produced; therefore, it is important to provide guidance to all design engineers before they start formulating the design solution. Figure 11-8 lists types of resources that are subject to standardization analysis.

   Parts.   One of the most common applications of resource standardization analysis is to parts, the basic building blocks of the system. Parts encompass nuts, bolts, screws, fasteners, capacitors, resistors, connectors, valves, lamps, bulbs, springs, plugs, switches, and any other item used in the system to create assemblies. The purpose of parts standardization is to produce a "shopping list" of parts that design engineers can use to select items for their design. This is not to establish a fixed limit on parts that can be used in a system. It is to require selected parts to be used wherever possible and limit requirements for nonstandard parts. Nonstandard parts are expensive to design, produce, and obtain through life.

   Current Parts Baseline.   The customer has an existing inventory of parts used to support systems currently in operation. This parts inventory can be used as a source of selecting parts for the new system. When a part from this list is selected, then the customer does not have to increase the number of parts in its supply inventory. A standardization analysis of existing inventory should produce a list of parts that are compatible with the technology baseline of the new system. The source for the current parts baseline should be identified in section 4 of the system requirements study. The list then is provided to each design engineer as an initial source of parts that can be used in the design.

   Parts Selection List.   There will be requirements for parts that are not in the existing customer inventory. When this occurs, design engineers should be pointed toward parts that conform to industry standards or specifications such as Institute of Electrical and Electronic Engineers (IEEE), Society of Automotive Engineers (SAE), International Standards Organization (ISO), and National Aeronautics and Space Administration (NASA) or, in the case of military programs, North Atlantic Treaty Organization (NATO) supply systems. Selection from one of these approved standards or specifications should enable the part to be purchased from several sources and also should ensure a consistent level of quality. As these additional parts are selected, their use should be recorded on a parts selection list that is distributed to all design engineers. This list becomes a second source for all design engineers to select parts for future applications.

Nonstandard Parts.   There always will be nonstandard parts in a new design. However, strict application of parts standardization should limit the number of nonstandard parts to the minimum feasible based on the technology of the design. Supportability engineering should assist design engineers in selecting appropriate parts through a continual dialogue as the system design evolves.

Tools.   Requirements for tools to support the system typically are created by the parts selected for the design. However, there can be another philosophical view taken toward tools, and that is using existing tools as the basis for selecting parts. For example, current maintenance personnel probably have a set of tools necessary to perform repairs on the system being replaced. A list of the tools in the set can be provided to design engineers with instructions to create a design that can be repaired using only these tools. On radically new designs, an even more creative requirement can be established, such as a requirement for design engineers to produce a system that requires no tools for disassembly and reassembly. This is feasible. If it becomes apparent that this dictatorial requirement cannot be achieved, then standardize the design so that it can be disassembled and reassembled using a single tool. Section 4 of the system requirements study (see Appendix B) should contain identification of tools that will be available to support the system when it is placed in service. Requirements for nonstandard tools are created by the design engineer. It is imperative that all efforts are made to eliminate nonstandard tool requirements from the design.

Support and Test Equipment.   The design technology baseline, operation concept, and support concept of a system generate requirements for support and test equipment. *Support and test equipment* can be loosely defined as anything that is not an integral part of a system that is necessary to operate or support the system. Support and test equipment typically is divided into the categories listed at Figure 11-9. Note that the generally, support and test equipment is divided into its use or technology. It includes items required for maintenance, operation, and training. Standardization analysis for support and test equipment is conducted in the same manner as was described earlier for tools. Support and test equipment tends to be more complex items than tools. This causes more expense in its procurement and support through life. Items of support and test equipment may be considered systems in their own right and require the same engineering activities

- Ground support equipment (GSE)
- Aerospace ground equipment (AGE)
- Maintenance support equipment (MSE)
- General-purpose electronic test equipment (GPETE)
- Special-purpose electronic test equipment (SPETE)
- Pneumatic support equipment
- Pneumatic test equipment
- Hydraulic support equipment
- Hydraulic test equipment
- Automatic test equipment (ATE)
- Factory test equipment (FTE)
- Materials handling equipment (MHE)
- Mobile STE
- Installed STE

FIGURE 11-9   Support and test equipment categories.

for their design and development as would a car, a ship, or an airplane. Therefore, every effort should be made to create a system design that makes use of support and test equipment that is already available in the user's inventory.

The system requirements study should list every item of support and test equipment that will be available when the new system enters service. In many instances, support and test equipment is shared with other systems; therefore, the capacity or utilization rate available for each item should be identified in the system requirements study. The only valid reason for nonstandard support and test equipment on a new design program is a technology baseline change that the user does not currently support.

Facilities.   A *facility* is defined generically as a place that provides a capability. This means that a facility could be a building, a confined space aboard a ship, a runway, or a vacant piece of real estate. There is no such thing as a standard facility. Figure 11-10 illustrates how facilities can be described. Since every facility is different, a standardization analysis investigates requirements for using facilities. The two issues that must be analyzed are capacity utilization rate and physical limitations. Section 4 of the system requirement study should list all facilities available to support the new system. Each available system must be studied by supportability engineering to determine its capacity utilization rate and its physical limitations to ensure that the new system will not exceed what is available.

Let's look at two examples: (1) capacity utilization rate and (2) physical limitations. The first example is a maintenance facility that repairs electronic assemblies for many different systems. Its total capacity is an output of 200 assemblies per month. The system being replaced generates 20 items per month that are repaired at this facility. Assuming that the electronic technology of the new system is very similar to that of the old system, then every effort should be made to generate no more than

By function or use:
- Operation
- Maintenance
- Support
- Training
- Storage

By type:
- Permanent
- Mobile
- General purpose
- Special
- Shared
- Dedicated

By features or construction:
- Building, mobile shelter, dock, tentage, pier, ground, prepared ground pad, shipboard compartment, aircraft compartment,
  - » Basic
  - » With specific capabilities
  - » With specific installed equipment

FIGURE 11-10   Facility categories.

20 items per month to be sent to this facility for repair. The second example is a maintenance facility that overhauls and refurbishes main battle tanks. In this program, the physical limitations of the maintenance facility may be critical. This analysis looks at the physical dimensions of doors to enter and exit, the load-bearing strength of the floor, the load-bearing strength of overhead cranes, and access space. If the new tank cannot fit through the doors or is too heavy for the floor or overhead cranes, then the existing facility is physically incapable of supporting the new tank. The existing facility would require extensive modification, or a new facility would have to be constructed. Either of these would greatly increase acquisition costs and through-life costs and therefore should be avoided.

Personnel and Training.   The areas of personnel and training may not be subjects that are commonly linked to standardization, but they always should be part of a system standardization analysis performed by supportability engineering. As discussed in Chapter 10, the cost of personnel required to operate and support a system constitutes a major portion of through-life costs. Therefore, any opportunity to limit this cost must be taken.

The system requirements study should identify the number of personnel and the labor hours of each type that will be available to operate and support the new system. Normally, system engineering focuses on operational personnel requirements, so supportability engineering must focus on support personnel requirements. Using the information on support personnel available to support the new system, an estimate of available maintenance labor hours can be developed that becomes a design-to goal for design engineering. The problem with this statistic is relating the overall goal to individual assemblies and items contained within the new system. The available maintenance labor hours can be allocated down the system architecture in a manner similar to allocation of mean time between failures (MTBF) and mean time to repair (MTTR), discussed in Chapter 5. MTBF can be used as a weighting factor in the apportionment, much like MTTR allocation. Figure 11-11 illustrates this technique using a complexity factor for each item to allocate labor hours. Care must be taken when using this technique because maintenance labor hour requirements will fluctuate with variations in system operating time. Therefore, the allocation should be based on maximum system usage rate during peak or surge conditions.

Personnel require skills and knowledge to perform assigned tasks. Knowledge is understanding, and skills are the ability to apply knowledge. Personnel receive the necessary skills and knowledge

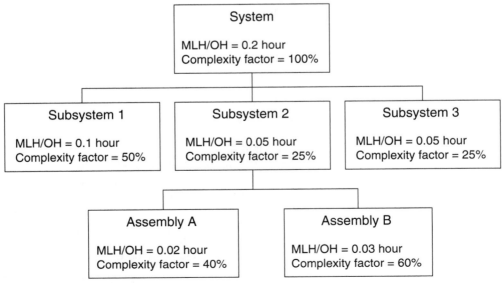

FIGURE 11-11   Maintenance labor hours allocation.

through training. There is an identifiable cost attributed to training of personnel. Personnel attend schools to obtain knowledge and then learn skills to apply the knowledge. It is desirable that existing available personnel assigned to support the system require the minimum possible training. Training can be measured in days of duration of training courses. A way of significantly limiting the duration of training for personnel to support the new system is to limit the volume of additional knowledge required. For example, existing maintenance personnel simply use built-in text (BIT) to test the old system and then replace the failed item. If the new system requires maintenance personnel to read schematic drawings and use general-purpose test equipment for troubleshooting, then they must receive extensive technical training to augment their knowledge base. The design of the new system created this additional training requirement. Thus another design-to goal can be established: Create a design that requires no additional knowledge for maintenance. A new system virtually always will require some type of skills training.

Packaging. Spares and support equipment require packaging for protection from damage during shipment, handling, and storage. Most industries have established standards for packaging materials and containers. Typically, there are several sizes and shapes of containers that are readily available. Use of these standard containers avoids the additional cost of custom-built containers. Additionally, there are industry standards for cleaning, preserving, and packing items. Any spare or item of support equipment should be capable of being packaged using standard containers and procedures—another design-to goal for design engineers. This is another area where supportability engineers must assist design engineers in attaining the goal. It is recommended that supportability engineering select a range of containers that become the "standard" for an acquisition program. Then design engineers can visualize the requirement and check their own design solutions. An unacceptable design would be one that will not fit into a standard container or require some special process to prepare it to be packed in the container.

Transportation. There are many different methods available to transport the system and the resources necessary to support it. Standardization analysis does not analyze these methods. The application of standardization analysis is to focus on the design characteristics of the system and resources that will allow them to be transported using standard methods. This design characteristic is transportability. The design of the system, each spare, and each item of support equipment must be accomplished so that it can be transported using standard methods. An example of an undesirable design would be one that is extremely fragile and requires special transportation. Another undesirable design would be one that is susceptible to damage from significant fluctuations in temperature, causing a requirement for transport by a method that has temperature controls. Neither of these designs had standard transportability characteristics.

Caution on Standardization Analysis

Application of standardization analysis has proven benefit in lowering the through-life cost of a system. However, there are some risks with standardization—obsolescence and finite quantities of limited resources. For example, a part selected to be standard for a design may be at the end of its manufacturing life. If the part is used in the design, it probably will not be available to purchase to make repairs. This must be a concern when selecting standard items. There is also sometimes a limit on the availability of resources. The design may be standardized to be tested on a specific item of test equipment. However, there are so many requirements for testing that the capacity of the test equipment is exceeded, causing a requirement to purchase an additional set of test equipment. The same analogy applies to facilities and personnel.

Technology Benefit Analysis

Standardization analysis seeks commonality and use of existing resources. But there is another side to new acquisition programs—application of new technologies. Technology improvements in the system design typically also provide opportunities to improve supportability. A technology benefit analysis looks for opportunities to apply state-of-the-art capabilities for support. This analysis looks at new and emerging technologies for application to the new system. Supportability engineering searches for how new technologies are being applied to other acquisition programs. It also looks to the future. There is a continual improvement and evolution of technologies. These may be in areas of reliability, maintainability, testability, transportation, support equipment, computer-based training, innovative materials, alternative methods of production, different power sources, or anything else. Performance of a technology benefit analysis requires access to information on new and future technologic innovations.

There is no set method of performing this analysis. It requires vision of what the rest of the world is doing and where technology is going in the future. Forty years ago, the idea that a system could test itself was considered a dream; today, BIT is standard. A technology benefit analysis looks into the future to the next beneficial technology improvement that can be applied to a new design. A thorough analysis typically identifies many technologies that have the potential to provide benefit to the new design; however, many of them may be in their formative stages and not be sufficiently mature to apply to the new design. The risk of using them may be too high to justify pursuing them. These are not eliminated from consideration. They are simply kept in abeyance for future consideration. A major system will have an operational life of many years. During that time, it will be upgraded and modified. Each time the system is changed, these technologies should be revisited at that time. The formal name for this is *preplanned product improvement* (P^3I). It is a controlled technology growth plan for a system over its useful life.

Support Infrastructure Analysis

The user has an existing infrastructure that supports all system in its inventory. This infrastructure consists of maintenance capabilities, supply operations, facilities, transportation, and personnel. A support infrastructure analysis investigates two different issues: responsiveness of the infrastructure to demands for support and any impact that introduction of the new system will have on the infrastructure and other systems it supports.

The user's support infrastructure should be analyzed to determine its responsiveness to requirements. The discussion in Chapter 9 highlighted the fact that one of the most significant negative factors that degrades operational availability is ALDT. ALDT is a reflection of the ability of the infrastructure to provide support on demand. This responsiveness is measured in both the time to respond to a requirement and the quantities of resources held awaiting a demand. The focus of this analysis is to identify the bottlenecks that cause delays or time wasted for no valid reason and how the infrastructure prioritizes its responses to demands. Figure 11-12 has been annotated with example times for the responsiveness of a support infrastructure. Visibility of these times allows analysis and detective work to determine ways to improve the overall responsiveness that will benefit the new system and all other systems supported by the infrastructure. Implementation of the results of this analysis should aid in improving operational availability for the new system and potentially all other systems supported by the same infrastructure.

Any infrastructure has maximum limits on its capacity to provide resources to all systems it supports. A typical example of this limit is the total quantity of fuel it is able to deliver over a specific time period. For example, both the existing system and the new system require diesel fuel. The infrastructure has demonstrated its capability to deliver all the fuel required for the existing system. However, the new system requires 100 percent more fuel than the existing system owing to a larger, more powerful engine. This increase in fuel requirement, when combined with the fuel requirements for all other diesel-powered systems, may exceed the maximum fuel-delivery capability of the infrastructure. The infrastructure must be augmented with a greater fuel-delivery capability. The problem increases in magnitude if other new systems currently being acquired also require more fuel. Supportability engineering must have visibility of the existing system, the new system, other systems in operation, and other systems being acquired simultaneously to quantity the complete problem so that steps can be taken to avoid it when the new system becomes operational.

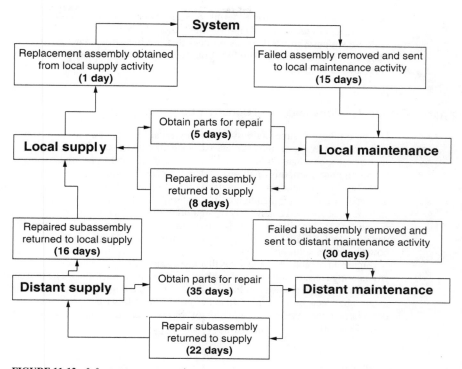

FIGURE 11-12   Infrastructure response times.

Introduction of the new system must be analyzed to determine any positive or negative effects this may cause. One of the obvious effects is exceeding the overall capacity of the infrastructure or requiring resources that are already dedicated to another system. However, there is another side to this issue. For example, the system being replaced uses 50 percent of the capacity of a maintenance facility, but the new system will not require use of the facility. This may look good for the new system, but in fact, it creates excess capacity at the facility that may have no other use. Other systems using the maintenance facility still require it. This actually will increase the through-life cost of the other systems. And it produces no real reduction in the user's total capability costs.

All three of these analyses are difficult to perform because they look outside the acquisition project. It takes diligence and persistence to perform these analyses. One of the hardest parts is obtaining creditable information to start the analysis. The user must be involved in identification of information sources. A well-prepared system requirements study should provide a reasonable starting point.

SUPPORTABILITY DESIGN CHARACTERISTICS

The functional supportability analyses described in this chapter provide supportability engineering with a methodology to identify critical issues related to the design, support, and cost of ownership of a new system. Using these analysis techniques enables supportability engineering to quantify issues in measurable terms or form boundaries for qualitative concerns. The analysis results lead to definition of significant design characteristics that have the highest potential to improve supportability. It is this process that produces the characteristics that were discussed in Chapter 6. An illustrative list of possible characteristics that may be developed for a system is provided in Figure 6-12.

Supportability Goals, Thresholds, and Constraints.

Design criteria for supportability eventually must be expressed in terms that are measurable. Many of the supportability design characteristics are clearly measurable when stated in terms of *no* or *must,* such as a design where all on-equipment maintenance can be performed using *no* tools. There is nothing ambiguous about this requirement. However, other requirement must be stated in statistics. Many of the statistics do not originate with supportability engineering. As discussed in previous chapters, the areas of reliability, maintainability, testability, safety, and human factors engineering contribute statistics that are used extensively by supportability engineering. One of the significant areas that supportability engineering is directly involved with is assessing the relationships between statistics developed by different disciplines and using the statistics in the context of supporting the system and how the design solution optimizes support.

Figure 11-13 indicates the areas that these statistics may address. Statistics used by supportability engineering are stated in terms of goals, thresholds, and constraints. The most common statistics used during acquisition are listed at Figure 11-14. At first, it may appear that there may not be a direct connection between some of these statistics and supportability design characteristics; however, they all contribute to the overall determination of the potential supportability of the system and its estimated cost of ownership.

Comparison Analysis Revisited

As supportability engineering develops goals, thresholds, and constraints, it is necessary to test them for reasonableness, adequacy, and compatibility. At the time when these statistics must be determined, the new system does not physically exist. Only a functional description is available. The only viable method of assessing the statistics is to return to the comparison analysis presented earlier in this chapter. The comparison system analysis provides a baseline against which these statistics can be tested. This is a classic "what if" scenario. By constructing an additional version of the comparison baseline, each of the statistics can be analyzed in terms of completeness and then be compared with the results of the system being replaced. This technique may reveal that some of the statistics have no effect on changing the outcome of the analysis, whereas others may have a dramatic effect. Use of the comparison analysis to match old to new provides a clear relationship between each statistic and its contribution to overall system supportability, operational availability, and cost of ownership.

Implementation

The efforts of supportability engineering will be wasted unless the results are implemented. Required supportability goals, thresholds, and constraints must be included in the product specification if one is to be written for acquisition. Then each design engineer must be tasked to ensure that his or her individual product incorporates applicable characteristics so that the system meets all supportability requirements. Typically, supportability engineers work with design engineers to advise and assist in incorporating requirements as a normal course of the design process. Additionally, supportability should be a major issue that is addressed at each design review, engineering test, and finally, product acceptance.

- System
    - Need oriented
    - Design-to or buy-to requirements
    - Maximum and minimum limitations
- Support infrastructure
    - Fixed basis
    - Variable implications
- Mission capability
    - Sortie rate
    - Force utilization rate
    - Response thresholds
- Operational availability
    - Peacetime
    - Wartime
    - Transition period
- Acquisition logistics
    - Design
    - Infrastructure
- Operational logistics
    - Supply
    - Maintenance
    - Training
- Design characteristics
    - Reliability
    - Maintainability
    - Testability
    - Human factors engineering
- Personnel requirements
    - Access time
    - Maintenance task time
- Support resources
    - Tools, SE & TE
    - Parts
- Operation resources
    - Fuel/power
    - Facilities
    - Training
- Infrastructure
    - Supply
    - Maintenance
    - Training
    - Transport
    - Facilities

FIGURE 11-13   Supportability engineering objectives.

Maintenance:

- Mean time to repair (MTTR)—weighted-average time required to return a system to an operable condition

- Mean restoration time (MRT)—weighted-average time required to return a software-based system to an operable condition

- Maintenance ratio—the ratio of maintenance done at each level of support

- Max time to repair (MAX_{TTR})—the maximum time allowable for any single maintenance action performed on the system

- Repair cycle time—the amount of time from when a failed item is removed from the system until it is repaired and returned to supply for use again

- Annual maintenance labor hours—the total number of labor hours required to support a system

- O&S cost per operating hour—the annual operation and support costs for a system divided by the number of annual operating hours

- Maintenance downtime—the total time a system is nonoperational due to corrective and preventive maintenance

- Waiting time nonmission-capable maintenance (NMCM)—the annual amount of time that a system is not mission capable due to corrective and preventive maintenance

- Percent organic support—the percent of total maintenance that is performed on the system by the operator/crew

- Maintenance hours to operating/flight hours—the ratio of hours required for maintenance to the number of operating hours

FIGURE 11-14   Supportability-related goals, thresholds, and constraints.

Functional Supportability Analysis Summary

The analysis techniques applied by supportability engineering ensure that every effort is taken to produce a physical design that can be supported efficiently and effectively. It has been stressed that all the significant work is done before the physical design is created. Timing of these activities is often much more important that the accuracy of results. If the activities are not performed during the conceptual formulation of the design, then they will not be effective. In some instances, the purpose of

Manpower and personnel:

- Crew size—the number of operators required for one system

- Labor cost per operating hour—the total annual cost of direct labor required to operate and maintain a system divided by annual operating hours

- Skill level limit—the maximum technical knowledge that an operator or maintainer can possess

- Maintenance hours by skill—the labor hours required for each type or classification of maintenance person

- Personnel costs/O&S—the ratio of personnel costs to total operating and support costs

Technical data:

- Technical document accuracy—the percent of accuracy of technical documentation delivered with the system necessary for operation and support

- Percent embedded TMs—the percent of technical documentation that is embedded in a software-based system

- TM correction rate—the percent of technical documentation that must be revised or corrected on an annual basis

Supply support:

- Waiting time—Non-mission-capable supply—the annual amount of time a system is nonmission capable due to awaiting resources.

- Parts availability—the percent of spares that are available when required

- Backorder rate—the percent of spares that are not readily available for replenishment

FIGURE 11-14   *(Continued).*

- Backorder duration—the average time required to receive a backorder spare

- Failure factor accuracy—the percent accuracy of predicted failure rates or MTBF

- Order-ship time—the average time from when an order for a spare is placed until it arrives at the correct destination

- Spares cost to TCO ratio—the ratio of cost of spares to overall cost of ownership

Support equipment & testing:

- On-system diagnostics—the percent of test diagnostics performed by BIT/BITE

- Diagnostics effectiveness—the percent of time BIT/BITE is actually required for fault detection and fault isolation

- Tools effectiveness—the percent of time that the correct tools are available when needed in terms of location and quantity

- Support equipment availability—operational availability for each item of support equipment

- Unit load support equipment—the quantity of support equipment that must accompany a system mobile or deployable when it moves to another location

Facilities:

- Facility utilization rate—the percent of the capacity of each facility that is actually used to support a system

- Facilities cost to TCO—the ratio of facilities cost to total cost of ownership

FIGURE 11-14 (*Continued*).

Training:

- Time to achieve proficiency—The amount of time required to train a person to a minimum level of proficiency

- Student failure percent—the percent of students that fail to complete a training course
- Embedded training—the percent of training course information that is imbedded in a software-based system

- Training costs per student—The average total cost to train one student to a minimum proficiency level

- Maintenance induced error rate—The percent of maintenance actions necessary due to damage during maintenance or improper performance of a maintenance action

- Training equipment availability—Operational availability for each item of training equipment

Overall system measurement options:

- Mission capability

- Operational availability

- Sustainability

- Cost of ownership (LCC, TLC, WLC)

FIGURE 11-14   (*Continued*).

the analysis is simply to be able to avoid negative outcomes. There are many cases where approximately 50 percent accuracy is very acceptable. This is sometimes the purpose of functional supportability analysis—attempting to limit the risk of making the wrong decision concerning supportability and cost of ownership.

Functional supportability analysis is a common activity that is applied routinely in virtually any industry where a complex system requires support for a long period of time. Proper application is proven to provide the ability to attain a balance among system performance, support requirements, and cost of ownership.

Specific cost issues:

- Money

  - Cost per operating hour
  - Cost per mission
  - Cost per maintenance action

- Personnel

  - Maintenance labor per operating hour
  - Maintenance labor per maintenance action

- Infrastructure

  - Fixed baseline
  - Variable rate thresholds
  - Threshold steps

FIGURE 11-14   (*Continued*).

PHYSICAL SUPPORTABILITY ANALYSIS

Supportability engineering is also responsible for identification of resources required to support the system being acquired. This activity is totally different from functional supportability analysis. It may be done at a much later stage of the acquisition process and may be performed by a completely different organization. The system design must be in its final stages of completion in order to perform physical supportability analysis. In some cases, several years have transpired since functional supportability analysis was completed, and in other cases, it may be only a few months later. Timing of performance depends on the status of the design evolution. Since subsystems and assemblies contained in the design of a system evolve at different rates, it is entirely possible for some physical supportability analysis to begin when other areas of a system are still under development. Every acquisition program is different, so timing of commencing physical supportability analysis depends on the specifics of the engineering schedule.

Physical supportability analysis is performed in three stages, as shown in Figure 11-15. The first stage consists of a series of activities that identify the complete range of resources that will be required to support the new system in its prestated operational environment. The second stage of

1. Identification of support resource requirements
2. Determination of optimal support infrastructure
3. Determination of resource quantities

FIGURE 11-15   Physical supportability engineering.

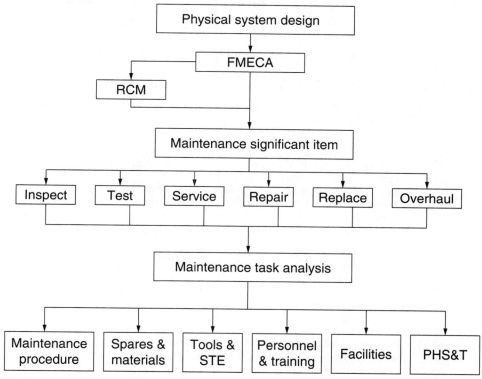

FIGURE 11-16   Resource identification process.

physical supportability analysis determines the optimal support infrastructure and maintenance policy to achieve a balance among performance, support, and cost of ownership. The third stage focuses on determining the minimum quantity of each resource that will be required to support the system at its prestated usage rate.

Resource Identification Process

The process for identification of support-resource requirements is a step-by-step approach that starts with the physical design and subjects it to a series of analyses that results in the definition of all resources needed to maintain the system in its operational environment. Figure 11-16 illustrates this complete process. As shown in this figure, identification starts with a physical system design. Typically, this consists of engineering design and assembly drawings with appropriate parts lists.

MAINTENANCE-SIGNIFICANT ITEMS

The major initial activity within the process is the identification of items that potentially require support. These items are termed *maintenance-significant items* (MSI), indicating that some type of maintenance support will be required. As shown at Figure 11-16, the failure modes, elements, and

---

- Test for unserviceability or failure

- Disassembly/reassembly without destruction

- Access to resource information

FIGURE 11-17   MSI selection criteria.

criticality anaylsis (FMECA) and reliability centered maintenance (RCM) analysis provide valuable input into this identification activity. The FMECA indicates that an item will fail and therefore that support will be required. This provides a basis for ensuring that for every way that a system can fail, there is a method of restoring it to an operable condition. Additionally, the RCM analysis identifies items that will require scheduled maintenance to preserve system operability. Both the FMECA and RCM analysis have been discussed in previous chapters of this text, so no additional discussion is presented here. There are some other criteria that also must be applied to identify an MSI. These criteria are listed at Figure 11-17.

It must be possible to identify that an item requires maintenance in order to be capable of supporting it. Typically, identification of this requirement is through some type of testing. The testing may be visual or manual or through automated testing or use of an item of test equipment. The testing method must be feasible to perform and reasonable to implement in the operational environment of the system.

The next criteriion for an MSI is that it can be disassembled and reassembled without destroying it during the maintenance process. This goes back to design criteria that were discussed in Chapter 6. Poor design techniques allow items to be welded or sealed in a manner that does not allow disassembly without actually rendering the item unusable.

Finally, sufficient information must be available to identify and document all the resources (i.e., tools, parts, etc.) that will be required to perform maintenance for the MSI. The information may be derived from assembly drawings, manufacturer source data, or other documentation produced during the design process or to be used for manufacture of the item.

Maintenance Task Identification

The next step in the process is to identify all maintenance tasks that potentially will be required to maintain each MSI. A *maintenance task* is an action performed to return the MSI to an operable condition. Maintenance tasks may include inspection, testing, removal, repair, replacement, servicing, or any other activity. An illustrative list of maintenance task descriptions is provided at Figure 11-18.

Maintenance Task Analysis

Maintenance task analysis (MTA) is the detailed, step-by-step analysis of a maintenance action to determine how it should be performed, who will be required to perform it, and all the physical resources necessary to complete the task. This is a very laborious and time-consuming activity, but it is extremely important. MTA results in identification of all the steps required to perform the task, spares and materials required by the task, tools and other support equipment needed, the person or persons who will perform the task, and any facility issues that must be considered to allow the task to be performed. When MTA has been performed for all MSIs, the results are a complete identification of all physical resources that must be available to support a system.

Access	Operate
Adjust	Overhaul
Align	Package/unpackage
Calibrate	Preserve
Clean	Process
Disassemble/assemble	Rebuild
Fault location	Remove
Inspect	Remove and replace
Install	Repair
Lubricate	Service
Mission profile change	Set up
Monitor	Test

FIGURE 11-18   Maintenance task descriptions.

Performing Maintenance Task Analysis

Performing an MTA starts with identification of each step of the maintenance task sequentially. Then each step of the task is analyzed to identify how it would be performed physically. After the physical description of how the task will be performed is completed, each resource necessary to support the task is identified. These resources include

1. Person or persons participating in each step, including a narrative description of what they are doing
2. Time duration of each person's participation
3. Tools or other support equipment needed
4. Parts and materials required for the step

After completion of these four activities, the results are analyzed to determine the following issues about the total maintenance task:

5. The total elapsed time for the entire task from start to completion
6. The type of person (or persons) required to perform the task based on their minimum technical capabilities, knowledge, and experience
7. Any additional training that must be provided to the person to ensure proper task performance
8. Any facility implications, such as space limitations, environmental controls, health hazards, or minimum capacity requirements

Finally, the MTA results must be analyzed to assess the item's compliance with all supportability issues, such as ease of maintenance, accessibility, standardization, etc., that may have been established by earlier functional analyses. The source for comparison of the physical support requirements for acceptability should be the design instructions issued to the responsible design engineer who created the design of the item. Any shortfalls or noncompliant features must be reported immediately back to the design organization for correction. This closes the loop between requirements for the design and the actual results of the design process.

The results of the MTA normally are documented in a common-source database that then can be used to quantify the support requirements for the system. The combination of the results of all MTAs produces the liability for support resources to support the system. This resource documentation then provides the information necessary to make the decisions as to how, where, and when support actually will be provided for the system when it is operating.

Performing Level-of-Repair Analysis

Level-of-repair analysis (LORA) is the most important physical supportability analysis business decision made during acquisition of a system. LORA produces the final answer as to how a system will be supported. LORA is performed in two steps: (1) using noneconomic decision criteria to make initial support decisions and (2) using an economic model to determine the most cost-effective alternative when there are no overriding reasons to provide support by one method over another.

The LORA process starts by identification of the options where maintenance can be performed. For example, it is common for systems to use either two or three levels of maintenance. Figure 11-4 illustrates a three-level support infrastructure where maintenance could be performed on the system at a local maintenance facility or at a distant maintenance support facility. This is a generic description of support capability. LORA produces a decision for each item within the system, indicating where each maintenance action for the item will be performed.

Noneconomic LORA decision criteria are a list of rules or guidelines that are used to determine if there is an overriding reason why maintenance should be performed. Figure 11-19 lists some typical examples of these criteria. Some organizations have policies that any item costing less than a predetermined price level will be arbitrarily discarded and replaced rather than being repaired. Technical feasibility or limitations for performing a repair may dictate a support solution.

Decisions that cannot be made using the noneconomic decision criteria are addressed using cost models that calculate the possible costs of all support options and then identify the least-cost solution. Figure 11-20 shows how all possible options for support of a system can be modeled to determine cost of each. Then the total cost of each option can be compared to determine the lowest option in terms of long-term support over the life of the system.

The LORA process produces the final support solution for the system. It determines where each required maintenance action will be performed, the physical resources that must be available to support performance of maintenance, and what the support infrastructure must be capable of sustaining throughout the operational life of the system. The results of LORA are documented and used as the basis for development of the physical resources required for support of the system.

- Mission requirements
- Safety
- Human factors engineering
- Constraints on existing support structure
- Special transportation factors
- Technical feasibility of repair
- Security
- Policy

FIGURE 11-19   Noneconomic LORA decision criteria.

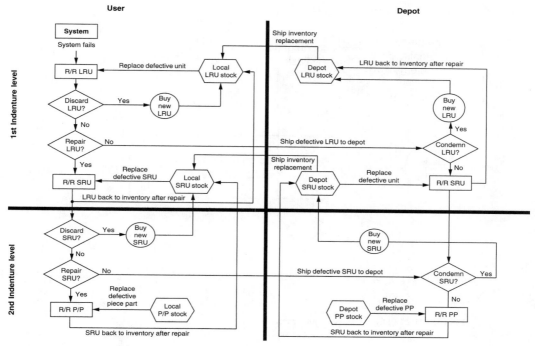

FIGURE 11-20   Economic LORA modeling.

Creating the Physical Support Resource Package

The physical supportability analysis process identifies and quantifies the liability to support a system in terms of the resources that must be available. Each physical resource required must be procured and positioned so that when it is required, it will be available. These resources are restated in Figure 11-21.

The physical supportability analysis process identifies the requirement for each of these resources as a straightforward technical activity; however, the procurement of sufficient quantities of each resource actually is a business decision because it requires the investment of money. Each resource must be analyzed to determine the reasonable quantity to be purchased based on the cost

- Spare parts and repair materials
- Trained personnel
- Facilities
- Technical documentation
- Tools, support, and test equipment

FIGURE 11-21   Physical support resources.

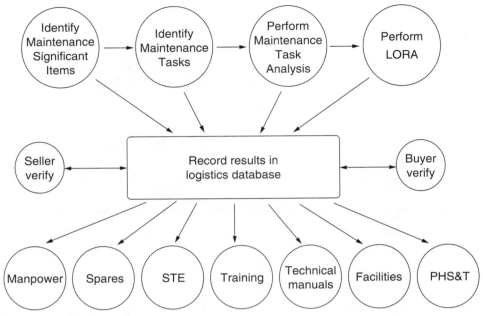

FIGURE 11-22   Physical supportability analysis.

of the investment. Figure 11-22 shows how the physical supportability analysis process provides this identification of resources and serves as a common source for development of the final resource package to support the system.

Spare Parts and Repair Materials

Typically, the repair of a system will require spare parts or repair materials. These must be available at the place the maintenance is to be performed; otherwise, the user will have to wait every time the system breaks. Waiting is detrimental to system availability, so it should be minimized by having some quantity of each spare part and repair material at the site. Common sense indicates that having excessive quantities is wasteful and expensive, so care must be taken to determine some minimum quantity, whereas at the same time not overinvesting and wasting money is permitted. Normally, the goal is to have sufficient materials on hand to respond to 95 percent of the requirements that may occur. There are many computer models that can be used to determine statistically the quantity of each item that should meet the 95 percent goal. The key figures of merit that most often drive this decision are the reliability of the item (either MTBF or failure rate) and the price of the item.

Personnel Requirements

The cost of personnel required to operate and maintain a system typically is one of the largest portions of cost of system ownership. The system engineering process includes efforts implementing design characteristics to minimize the requirements for personnel. The physical supportability analysis process provides a method of verifying that the systems engineering efforts have been successful. Reliability characteristics indicate the frequency that a support action will need to be performed, and maintainability characteristics indicate the duration of the support action. The combination of these two figures of merit provides an estimate of the personnel requirements for system support.

Facility Requirements

As stated previously, the cost of facilities tends to be another very significant portion of system cost of ownership. The broad category of facilities includes operation, maintenance, training, storage, and any other places necessary to support the system. The physical supportability analysis process provides identification of facility utilization rates and minimum capacity requirements. This analysis should focus on using existing facilities to the maximum extent possible. Since facilities typically are shared with other systems, it is important to ensure that the new system being developed does not exceed its allocated share of existing facilities, which would create a requirement to obtain a completely new facility.

Technical Documentation Requirements

The user of the system will require a technical documentation that provides information on proper system operation and maintenance. The information must be complete, unambiguous, and 100 percent technically correct. At the same time, the information presentation should be as simple and easy to use as possible. The physical supportability analysis process produces a clear description of everything the user must be capable of doing to support the system. This information, which has been documented accurately in a logistics database should serve as the sole input source for preparation of any technical documentation to be delivered with the system.

Support Equipment Requirements

The design of the system dictates requirements for support equipment, test equipment, and tools. The systems engineering process strives to minimize these requirements through proper design characteristics and the standardization activities discussed earlier. The physical supportability analysis process results in identification of the support equipment that will be needed to operate and maintain the system. This can be viewed as an assessment of the goodness of the results of the systems engineering process and its attainment of a design that requires the minimum support equipment. The MTA results that have been documented in a logistics database should serve as the single source for quantification of support equipment requirements for the system.

PHYSICAL SUPPORTABILITY ANALYSIS PROCESS

The physical supportability analysis process is applied as the functional design is transformed into a physical design. The process provides an audit trail of all decisions made to produce the final support capability for the system. The process follows the design activities performed in windows D4 and D5 (see Figure 2-8) of a design program, and then the results are validated during window D6 prior to delivery. Figure 9-23 shows the relationship between the design process and the physical supportability analysis process. Evolution of the design dictates when physical supportability analysis can be done. The process starts with identification of all possible MSIs. The list of possible MSIs is reduced by applying the noneconomic LORA criteria to eliminate any item where an overriding issue determines its support solution. For the items remaining on the MSI list, the next step is to identify all possible support actions that may be required to support that item.

The noneconomic LORA criteria are applied again, this time to eliminate any maintenance task from further consideration. Elimination may be for various reasons, but typically it might be simply that it is not technically feasible to perform the task, or it may be more advantageous to discard an item and purchase a replacement rather than performing a repair. The MSIs and maintenance tasks that have no resolution using the noneconomic LORA criteria then can be resolved using cost-based LORA modeling. Modeling required input data, so a very simple and quick maintenance task analysis is performed to create the minimum data set required to run the selected LORA model. Typically, the minimum input data consist of item price, MTBF, MTTR, and requirements for special facilities,

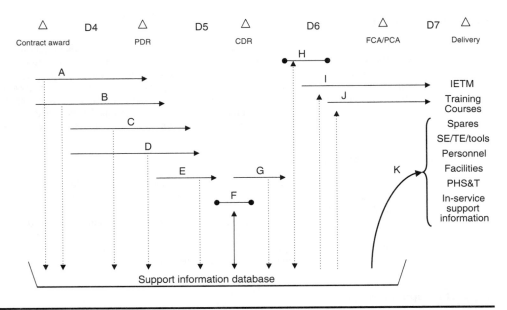

FIGURE 11-23   Developing the physical support solution.

- A–Identify all potential maintenance significant items that may require physical support
- B–Apply noneconomic LORA Criteria to eliminate any items from further consideration
- C–Identify all potential maintenance tasks for remaining MSI's
- D–Apply noneconomic LORA Criteria to eliminate any maintenance tasks from further considerations
- E–Perform limited maintenance task analysis to develop the minimum data set required for a LORA model
- F–Use LORA model to produce final support decision
- G–Perform detailed maintenance task analysis to develop sufficient data to identify final deliverable products
- H–Validate the results of step G by performing all maintenance actions on the prototype system

Note: The results of steps A through H are documented in a support information database

- I–Use the results of validation as the sole source for development of maintenance manuals
- J–Use the results of validation to produce a training needs analysis as the basis for training course development
- K–Use the support information database as the common source database for development and delivery of the physical support solution

special test equipment, and other personnel-related issues. LORA modeling should produce the final method-of-support answer for each MSI in the system. Then a detailed maintenance task analysis is carried out to identify all resources that will be required to support the system.

As can be seen in Figure 11-23, the results of all these steps are recorded in a support information database. The information may be documented using GEIA-STD-0007, Logistics Product Data, as the basis for creating standard information that can be shared through product life cycle support (PLCS). It is important that the support information be validated before it is used as the source of the final deliverable physical support package. Thus the maintenance task analysis results are validated by actually performing each task on the first representative system. This may be either the prototype that is being built for qualification testing or the first production system prior to first-article testing. The validated data then are used as source data for development of interactive electronic technical manuals (IETMs), other maintenance manuals and related documents, training needs analysis for training course development, and development of all other physical resources. This physical supportability analysis process has proven to be the best method for creating the final support solution for a system.

The supportability analysis techniques presented in this chapter are proven methods to improve system supportability and lower the cost of ownership. Proper application during the acquisition of a system ensures that the user will have adequate support capability for a minimum investment in support resources. Functional supportability analysis must be an integral part of the systems engineering process to be successful, and the physical supportability analysis process should validate this success.

CHAPTER 12
CONFIGURATION MANAGEMENT AND CONTROL

Configuration management   A process for establishing and maintaining consistency of a product's performance, functional, and physical attributes with its requirements, design, and operational information throughout its life.

The activities discussed in previous chapters described how the design of a product evolves from concept to reality. This evolution must be documented and controlled. *Configuration management* is the process that maintains total visibility of this design evolution. This is achieved through a series of detailed events that provide complete documentation of each change to the product design. The importance of configuration management can be summarized in three words: *interoperability, supportability,* and *reproducibility*. Figure 12-1 illustrates that configuration management is the foundation for system success.

INTEROPERABILITY

Products must operate with other products, mate with power sources, use support resources, and fit into higher systems. *Interoperability* is the highest requirement for configuration management. This can be achieved only by defining and controlling the physical and functional interfaces between the product being developed and all other products or systems where it will be used, installed, or supported. The system architecting and system engineering activities discussed in Chapter 4 must incorporate all interoperability requirements in the final design of the product. Configuration management is responsible for maintaining the documentation of any interface requirements for the product. Typically, this interface is documented in an interface control drawing (ICD) where necessary, or the interface may be identified through established industry or organizational specifications, such as those published by the American National Standards Institute (ANSI) or Institute of Electrical and Electronics Engineers (IEEE). Failure of the product to adhere to the necessary interfaces will result in it being incapable of operating or being supported.

Figure 12-2 provides some examples of interoperability. The first example illustrates that the owner of the automobile will purchase fuel from many different sources, so the automobile must be capable of accepting fuel from a standard fuel-delivery system. The automobile and the fuel-delivery system interface at the point where the refueling pump physically is inserted. The diameter of the opening in the fuel receptacle of the automobile must allow the pump handle spout to fit appropriately to accept the fuel flow.

The second example shows that a television must be designed to receive a standard transmission signal. A television produced for use in the United States will not work if used in the United Kingdom because the standards for transmission signals are different. The third example is similar to automobile refueling but much more critical. Aircraft refueling requires a pressure seal, so the refueling

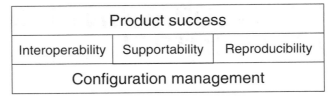

FIGURE 12-1 Basis for configuration management.

Product	Interoperability Requirement
Automobile	Refueling pump
Television	Transmission signal
Aircraft	Refueling nozzle
Missile	Aircraft
Aircraft	Air traffic control
Refrigerator	Electrical outlet
Wireless telephone	Communications network

FIGURE 12-2 Interoperability examples.

nozzle must be capable of achieving an exact fit to the aircraft's refueling receptacle. The next is an example of two different products, a missile and an aircraft, that must work together to perform. The mechanical mounting, electrical connection, and data link where the two are joined must be predefined before either is designed. Both the aircraft design organization and the missile design organization must have this predefined interface before each product design starts. This is a good example of where an ICD is mandatory. An aircraft must be able to communicate with applicable air traffic control systems. International civil aviation agencies have established standards that proscribe radio frequencies that are used for communications. Any aircraft certified as being airworthy must be capable of meeting these standards. The consumer who purchases a new refrigerator expects to be able to take it home and plug it into a common electrical receptacle, and the consumer who purchases a wireless telephone expects that it will work with the local communications network. Both these consumer products are designed to interface with predetermined sources. Each of these are examples of the importance of interoperability where products would not be successful if the necessary interfaces were not achieved.

SUPPORTABILITY

The requirement for configuration management to achieve a *supportable* product has been illustrated many times in previous chapters. For support to be planned and provided adequately, the product configuration must be completely documented and controlled. The ability for physical support resources

Product	Supportability Criteria
Aircraft	Operation manuals
Printer	Replacement ink cartridge
Automobile	Replacement oil filter
Engine	Metric hand tools
Circuit card	Electronic test equipment

FIGURE 12-3   Supportability examples.

to be identified, procured, and used is based on the premise that the product, when used, will be consistent with the design produced. When the product fails during use, any parts required for repair must fit. Figure 12-3 lists examples of supportability concerns that are met through good configuration management practices. Pilots continually refer to operation manuals for the aircraft they fly. The manuals must be technically correct and be identical to the configuration of the airplay they are flying. The replacement ink cartridge for a printer must fit. Many companies manufacture oil filters for cars. The manufacturer must know the exact interface measurements and performance characteristics required for its filter to fit mechanically and operate properly. There are two common measurements of hand tools, Imperial (inches) and metric (millimeters). A mechanic must have the correct type in order to work on an engine. The repair of a circuit card requires fault diagnosis before the failed component can be replaced. Therefore, the correct mechanical and electrical connections must be defined and controlled so that the appropriate test equipment can be purchased or designed. Any change to the design of the circuit card must be analyzed to determine if a change is also required for the test equipment. Supportability engineering relies on configuration management to maintain accurate records of the design of the product so that required support can be created and sustained.

REPRODUCIBILITY

The third justification for configuration management is *reproducibility*—the ability to produce an item identical to one produced previously. This includes the system and any subsystems, spares, and structures. Reproducibility allows manufacturers to continually produce items that are physically and functionally identical throughout the manufacturing process and then, at some point after the initial production ceases, to restart an additional production run. Only through sound configuration management practices can this be achieved.

CONFIGURATION BASELINES

The configuration of a system is managed through the establishment of baseline descriptions. These baselines control the development of the design from concept to production. The baselines of a system reflect its characteristics. They become the focal point for all activities related to the system. The

- *Functional baseline*—Quantitative performance parameters and characteristics and design constraints of a system including operational and support parameters and their respective tolerances

- *Allocated baseline*—Quantitative performance parameters and characteristics and design constraints for individual items which will comprise a system including operational and support parameters and their respective tolerances

- *Product baseline*—The approved, documented physical configuration of an item which, when manufactured to approved standards, has been tested to achieve the parameters and characteristics of the functional baseline

FIGURE 12-4   Configuration baselines.

most common names for these are shown at Figure 12-4. System baselines can be related to the documentation necessary to record functional and physical attributes. They form the overall control for evolution of the design.

Functional Baseline

The system architecting and system engineering processes discussed in Chapter 4 develop the functional requirements for a system. These requirements include measurable performance capabilities and tolerances, environmental limitations, supportability characteristics, reliability, maintainability and testability statistics, materials, and acceptance criteria. The functional baseline does not describe the physical system; it describes what it can do, where it can be used, and how it is to be tested to ensure that these capabilities are achieved. The functional baseline for a system normally is recorded in a single document. This document has several names, such as *product specification, technical specification, performance specification, engineering specification,* and *functional specification.* The functional baseline is the overarching parameters of the system capability. The functional baseline eventually will be used to assess any product design for its acceptability and suitability to meet the user's needs. It is the connection between the user's need that was discussed in Chapter 3 and the final design of the system that is delivered to the user to meet the need.

Allocated Baseline

The design evolves from the top-level specification, which describes the total system functionality, down to the functionality of each item, both hardware and software, that must be contained within the system. This is created following the functional block diagram of the system described in Chapter 4. Systems engineering decides how each function of the system block diagram will be achieved in the design solution, thereby allocating each function down to the items that will meet the requirement. Creation of the allocated baseline is an interim step in the design process to ensure that the final solution meets the functional requirements of the specification.

Product Baseline

The final baseline describes the actual, measurable performance and supportability characteristics of a physical system. This is a generic baseline. It is used to quantify the minimum capabilities of any system that has been manufactured and tested to achieve the requirements delineated in the functional baseline. The manufacturer of a system uses the product baseline as the primary description of the overall functional capabilities of an item that has been produced. The buyer of a system uses the product baseline to quantify the capabilities of the system when placed in its intended operational environment. All future changes to the design of a system, once production commences, are made against the product baseline.

CONFIGURATION ITEMS

The complete configuration of a system consists of subsystems, assemblies, components, parts, and software that are combined to perform a specific function. This assemblage creates the overall baseline of the system. Configuration management controls the system baseline through identification, documentation, and recording of the status of the individual items at each level of the system architecture. Any item, hardware or software, that forms a part of the system that has significant participation in system functionality is designated a *configuration item* (CI). Typically, systems contain hardware configuration items (HWCIs) and computer software configuration items (CSCIs).

The parameters that are controlled typically are referred to as the *form, fit,* and *function* of the CI. The form of a CI is its physical properties, such as size, weight, and subordinate contents. The fit of an item is its physical and functional interface with its next-higher assembly. And its function is a technical measurable description of what it does.

Configuration Identification

Configuration management establishes control of the design of a system through the documents required to establish the baseline for the system and all CIs contained therein. Configuration management issues and controls the numbers and other identifiers affixed to the items and documents such as part numbers and specification numbers. Where necessary, serial numbers, lot numbers, and revision numbers are also used to document versions or variations of CIs. The identification of an item includes the approved documents that identify and define the item's functional and physical characteristics in the form of specifications, drawings, associated lists, interface control documents and related specifications, process methods, and related documents.

CI Documentation

A series of documents is used to control the form, fit, and function of a CI. Figure 12-5 lists the documents required to control HWCIs. These documents, when consolidated, establish the physical relationship (sometimes called the *parent-child relationship*) of each individual CI; identify each

- Engineering family tree drawing
- Assembly drawings
- Parts list
- Acceptance test
- Process control drawings
- Release record

FIGURE 12-5   HWCI documentation.

- Sequential list number
- Manufacturer of the part (name or organization identification number)
- Part number or identification number
- Technical name of the part
- Quantity of the part used in the assembly
- Cross referencing number to the assembly drawing
- Any specification to which the part must be qualified

FIGURE 12-6   Parts list contents.

subordinate CI; the contents of each CI; the test procedures required to verify that a CI meets its form, fit, and function tolerances; any critical manufacturing processes for a CI; and identification of the most current change status of each CI. The specific set of these documents required to control the configuration of a CI may be referred to as its *technical data package.*

The engineering family tree drawing identifies the relationship of all CIs by depicting where they fit into the overall system architecture. This visual representation of the CIs contained in a system provides a valuable reference for all organizations participating in the project.

The assembly drawings for a CI illustrate how it is physically assembled, with identification of each part or material that is required. This drawing is the final product of design engineering. It is used as the basis for manufacturing the CI, and it is also used by supportability engineering to determine support resources.

The parts list is a tabulation of all subordinate CIs, parts, and materials identified on the corresponding assembly drawing. Figure 12-6 lists the minimum technical contents of a standard parts list. The parts list also identifies interconnecting items required for interface and attaching parts required for installation of the CI.

The acceptance test describes the testing process, procedures, and equipment that are used to verify that the item meets its functional requirements. The acceptance test procedure is used by the manufacturer to prove that a CI that has been produced by its manufacturing activity meets its form, fit, and function specification requirements. The acceptance test procedure links the CI back to its original design parameters. It is extremely important to ensure CI interoperability, supportability, and reproducibility. It validates interchangeability of identical part-numbered items.

In some cases, the manufacture of an item may require a special or unique process. If this is applicable to a CI, the process control drawing describes the step-by-step events that must be followed during manufacture. Figure 12-7 lists typical types of process control drawings.

- Soldering standard
- Component preparation method
- Lathe setup procedure
- Cable wiring loom setup
- Conformal coating application
- Packaging procedure
- Component cleaning procedure
- Painting procedure

FIGURE 12-7   Process control drawings.

- Document identification number
- Date of initial release from engineering
- System or project
- Subsequent approved changes
- Date of subsequent change
- Document (engineering change order) authorizing change
- Effectivity of the change

FIGURE 12-8   Release record information.

The final category of configuration documentation is the release record. This document is the controlling mechanism for determining the approved configuration for each CI. The term *release* refers to a document being approved for release from engineering to all other organizations for use in procurement, manufacturing, and support. The engineering process discussed in Chapter 4 creates the design of a CI. This process includes the initial design of the CI and all subsequent revisions, changes, and modifications. Internally, engineering organizations establish some type of informal control of items as they progress through the design process. Designs are in constant change. It is only when the design of a CI has been *released* for use outside the engineering organization that the configuration must be rigidly controlled. Release records must, at a minimum, include the information listed in Figure 12-8. Configuration management maintains a release record for each CI.

SOFTWARE CONFIGURATIONS

Software is an integral part of many systems, and its configuration also must be controlled. Since software is never translated into a physical item, it always remains functional. Thus the documentation required to control the configuration of software is different from that discussed earlier for hardware. The typical names for software configuration documentation are given in Figure 12-9.

Software Performance Specification (SPS)

The software performance specification describes the functional capabilities of the total software contents of a system. It is the basis used by software engineering to identify, develop, and control all lower-level software modules required to meet the system performance specification discussed in Chapter 4.

- Software performance specification (SPS)
- Top-level computer software configuration (TLCSC)
- Lower level computer software configuration (LLCSC)
- Computer software configuration item (CSCI)
- Unit development folder (UDF)
- Software development test procedure (SDTP)
- Release record

FIGURE 12-9   Software configuration documentation.

Top-Level Computer Software Configuration (TLCSC)

The actual functional content of the system software architecture is contained in the top-level computer software configuration document. It identifies each high-level functional segment and the inputs and outputs between the segments. An example of this document would be that of a word-processing program. A word-processing program allows creating a new document, editing an existing document, filing the document, spell checking the document, and formatting. Each of these is a high-level function of the software program.

Lower-Level Computer Software Configuration (LLCSC)

The lower-level computer software configuration document identifies each subfunction required to perform a specific function within a specific high-level function. It could be easily described as the child of a parent TLCSC. It also identifies the inputs and outputs of software components within the LLCSC plus inputs provided to and received from other LLCSCs. Both the TLCSC and the LLCSC document the results of the software systems engineering process.

Computer Software Configuration Item (CSCI)

The actual software code is contained in the CSCI document for each module. This document also identifies the inputs to the module, the functions performed by the module, and the outputs created as a product of exercising the software. This documents the result of the software design engineering process. It is the focal point for development and maintenance of software.

Unit Development Folder (UDF)

The hardware design process creates assembly drawings that are used to manufacture the physical item. However, software is never translated into a physical item; it always remains functional. Therefore, the basis for its development must be formally preserved for quality testing and future modifications. The UDF is controlled and managed throughout the life of the system where the software is installed.

Software Development Test Procedure (SDTP)

The software must be tested to ensure that it works. This may be done using a software test facility rather waiting until it is actually installed in the system. The system acceptance test procedure assesses the functionality of the combination of hardware and software. Thus a stand-alone test procedure is required for control of testing prior to installation. This is the purpose of the software development test procedure. This document contains the specific functional testing that is performed to determine if the software is technically and functionally compliant with the SPS and the system performance specification.

Release Record

The release record for a CSCI is very similar to that described previously for hardware. In addition to the information contained in the hardware release record, the software release record also contains identification of revisions (revision a, etc.) and specific links to LLCSC and TLCSC applicability. As with hardware, the software release record is the basis for configuration control and reflects the latest version of each CSCI plus any previous versions.

SYSTEM CONFIGURATIONS

The system baselines, discussed earlier, define documentation created by a series of engineering activities. The baselines provide descriptions for the system. The developer or manufacturer uses the baselines as the basis to create physical products by relating CIs to the established baselines. The amalgamation of physical items is described as various system configurations. A list of typical names of system configurations is provided in Figure 12-10. It may seem cumbersome having configurations named with the prefix *As-*; however, it is appropriate because the name states that a physical item conforms to the form, fit, and function of an applicable baseline *as documented*.

As-Specified Configuration

The as-specified configuration of a CI is a definition of its form, fit, and function. Normally, this configuration applies to the item that is located at the top of the engineering family tree drawing. Initially, the performance specification document is the single definition of this configuration. As the design progresses through the engineering process, the as-specified configuration expands to encompass all lower-level specifications for each CI.

As-Designed Configuration

The as-designed configuration includes all released documentation for the CI. This is sometimes called the *product baseline configuration.* The as-designed configuration is the basis for all procurement and manufacturing activities. It depicts and describes the generic configuration of an item and any applicable manufacturing processes required to ensure that it meets specified tolerances. The as-designed configuration is the "paper design" of an item.

As-Manufactured Configuration

A manufacturing organization receives the as-designed (paper design) configuration and produces the physical item. Theoretically, the as-manufactured (physical) configuration is identical to the as-designed (paper) design. In reality, minor differences may occur owing to differences in the course of procurement and manufacturing. These differences are described as either *deviations* or *waivers.*

A *deviation* is a minor difference that has been approved prior to the start of manufacturing a specific item. For example, the parts list for a CI may require a common part such as a resistor from a certain supplier. The purchasing department may determine that the supplier cannot meet the delivery schedule for that part to have it available when it is required for manufacture of the CI. The department identifies an alternate source that produces an identical form, fit, and functional item and can deliver the required part. Using the alternate part has no effect on form, fit, and function of the CI.

- As-specified configuration
- As-designed configuration
- As-manufactured configuration
- As-delivered configuration
- As-maintained configuration

FIGURE 12-10   System configuration.

A one-time deviation could be used to authorize using the alternate part until the original part is available. The deviation must be documented in the CI release record.

The other category of minor difference is a waiver. A *waiver* is approval of a difference after an item has been manufactured, and it is a one-time difference for a specific item. For example, during the manufacture of a bracket, a hole is drilled in the wrong location. When the error is realized, the hole is redrilled in the correct location, and the wrong hole is repaired. It has no effect on form, fit, or function. Manufacturing may request and receive approval for a waiver to deviate from the exact configuration documentation requirements to allow use of this repaired item. The waiver is also documented in the CI release record.

The incorporation of deviations and waivers results in an as-manufactured configuration that meets the form, fit, and function requirements of the as-designed configuration but is not physically exactly the same. Configuration management is actively involved with procurement and manufacturing activities to maintain control of the actual configuration of an item that is produced.

As-Delivered Configuration

The as-delivered configuration is physically identical to the as-manufactured configuration. The difference between the two is that the as-delivered configuration is created for each individual system, whereas the as-manufactured configuration applies generically to all like items. The as-delivered configuration also includes all software that is installed on the system. The purpose of the as-delivered configuration is to document the specific CI content of a system by serial number or lot number. For example, the as-delivered configuration of a system would identify that serial number 0005 module is located in serial number BB024 assembly that goes into subsystem serial number A161 that is in system number 12. Each CSCI installed in the delivered system is documented by revision number. This is the complete parent-child relationship from lowest CI to the top-level delivered system, or its as-delivered engineering family tree drawing. Both the automotive and aviation industries routinely maintain an as-delivered configuration log for all products sold to users.

As-Maintained Configuration

After a system is delivered to the user, there are changes to its as-delivered configuration through normal maintenance actions. When a CI fails, it is replaced by an identical item; however, the replacement item will have a different serial number. Therefore, the individual configuration of the system has changed, and the records pertaining to that system must be updated to reflect this new configuration. This updating provides a record of the most current configuration of the system. Occasionally, a design flaw is identified after several systems have been delivered. A retrofit kit may be produced to correct the design flaw in delivered systems. The application of the retrofit kit to each system must be documented in the as-maintained configuration of the system. A good example of the use of a retrofit kit comes from the automotive industry, where there is a recall for specific delivered automobiles. Notification is sent to owners of affected automobiles based on the as-delivered configuration maintained by the automobile company. The owner of each automobile takes it to an authorized service point, where the recall retrofit kit is installed. The service point reports completion of retrofit kit installation to the company, which, in turn, updates the as-maintained configuration for that specific automobile.

CONFIGURATION AUDITS

Three configuration audits are performed to ensure that the configuration of a system has been controlled accurately throughout its design and manufacture: a functional configuration audit, a physical configuration audit, and a production quality audit. Each audit has a specific purpose to ensure that all configuration documentation is correct and that the system design meets its form, fit, and function requirements.

Functional Configuration Audit (FCA)

The FCA is conducted after the system has successfully completed all performance testing. The performance testing may include field trials, flight testing, qualification tests, or some other type of testing that is deemed necessary to demonstrate all functional characteristics of the as-designed configuration. At completion of performance testing, the results of the tests are compared with the form, fit, and function requirements contained in the system performance specification. The purpose of the FCA is to demonstrate that the as-designed configuration of the system meets all specification requirements.

Physical Configuration Audit (PCA)

After a system has successfully passed the FCA, the next event is to verify that all configuration identification documentation for the as-design configuration is correct. This is done through a PCA, where every applicable document listed in Figures 12-5 and 12-9 is compared with the physical system. In some cases, the system that successfully passed the FCA is disassembled so that all internal CIs can be checked. This is sometimes called a *tear-down audit*. The purpose of the PCA is to make sure that the configuration identification documentation that is used for procurement and manufacture accurately reflects the design of the system that successfully passed functional testing. Both the FCA and the PCA are key events that occur only after the design is completed but before procurement and manufacture can commence.

Production Quality Audit (PQA)

A PQA is performed on each system at the completion of all manufacturing and acceptance tests that have produced the as-manufactured configuration. The purpose of the PQA is to establish the as-delivered configuration of each individual system. Serial numbers or lot number of physical items and version or revision numbers of software programs are recorded. As-delivered configuration logs, or registers, are created for each delivered system. Maintenance of these logs allows the company to monitor system success, provide a basis to problem investigation, and initiate design changes where required.

CONFIGURATION CONTROL

Formal configuration control starts with official release of configuration identification documents. As discussed previously, this release is documented in the document release record. After release, changes can be made to the released document only using a formal configuration control process that is administered by configuration management. Any change must be documented to provide an audit trail for all decisions and approvals that affect form, fit, or function. Every organization that implements configuration management has a name for the form that is used to initiate a change. It may be called an *engineering change request* (ECR). The originator of the ECR describes the change and identifies all documents that may be affected by the change.

Figure 12-11 illustrates the change control process. The ECR is submitted by the originator to configuration management, where it is assigned a control number. It is then passed to the engineer responsible for the documents possibly affected by the change. The engineer comments on the recommended change and returns it to configuration management. The engineer cannot approve or disapprove the change, only provide comments. Configuration management then forwards the ECR to the configuration control board (CCB). Figure 12-12 shows the typical membership of a CCB. The CCB is responsible for analyzing the change and then making the final decision, based on the engineer's comments and their own analysis, as to whether the ECR should be approved or disapproved. If the ECR is approved, it is returned to configuration management, which prepares and issues an engineering change order (ECO). The ECO is the authority for changes to configuration identification

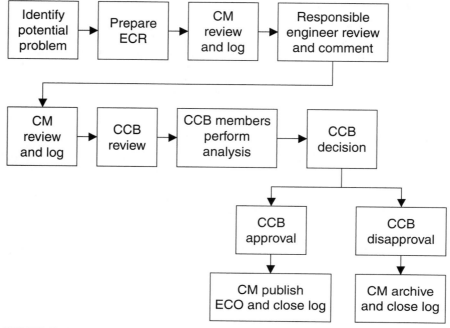

FIGURE 12-11   Change management process.

- Systems engineering
- Electronic engineering
- Mechanical engineering
- Software design
- Human factors engineer
- Safety engineer
- Production or manufacture engineering
- Supportability engineering
- RM and T engineering
- Quality assurance

FIGURE 12-12   Configuration control board.

- Operational availability
- Reliability
- Maintainability
- Testability
- Standardization
- Cost of ownership
- Spares
- Technical documentation
- Training
- Support equipment
- Personnel
- Safety
- Human factors

FIGURE 12-13   Supportability analysis of changes.

documentation. The release of the ECO is recorded on each affected CI's release record. ECRs that are disapproved are returned to configuration management, where the action is closed in their records.

Any change must be analyzed for its effect on supportability. The analysis should include the issues listed at Figure 12-13. Incorporation of a change may result in degradation of supportability characteristics of the system, added support costs, or both. Supportability engineering must be a voting member of the CCB so that any adverse impact of a recommended change can be identified and analyzed thoroughly prior to the change being approved.

Configuration Status Accounting

Configuration management uses a configuration status accounting (CSA) system to record and control all configuration identification documentation, create system configurations, change processing, record change implementation, and record changes to delivered systems owing to maintenance actions and retrofit. Figure 12-14 depicts the architecture of a CSA system.

The document information segment of the CSA records the released status of documentation for hardware CIs, documentation for software CIs, and all specifications and controlled processes. It also establishes a linkage between this documentation in a CI structure from which the as-designed system configuration is created. The CSA maintains a history of each document.

The change processing segment registers engineering change requests, records the configuration control board approval or disapproval decisions, and contains any engineering change orders that result from change approval. This is a very important segment because it gives visibility to all ongoing change activity. In many cases, a change will be approved, but it will not be implemented immediately for various reasons. Change effectivity may be delayed for the next production lot or to group changes for efficiency of implementation.

The change implementation segment records the actual implementation of changes as they are incorporated. All effects of an approved change must be identified on the engineering change order. Changes may affect the system hardware and software design and delivered support resources, such as spares, test equipment, and technical documentation. This segment also records the development of retrofit kits to incorporate the change into delivered systems.

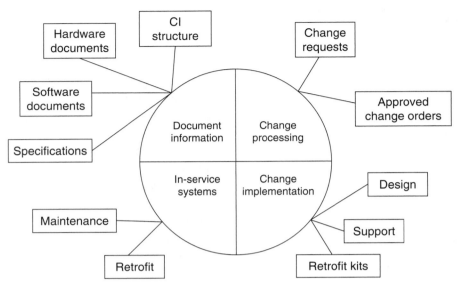

FIGURE 12-14 CSA system.

The fourth segment of the CSA system tracks all configuration changes to delivered systems. These changes are due to normal maintenance actions, where serial numbered CIs are exchanged to repair a system. The installation of retrofit kits is recorded. This segment allows visibility of the as-maintained configuration of each delivered system.

Configuration management, as has been discussed in this chapter, is an extremely important issue. It provides control of the design of a system from concept through in-service. Configuration management ensures system interoperability, supportability, and reproducibility.

CHAPTER 13
SUPPORTABILITY ASSESSMENT AND TESTING

Supportability must be assessed and tested throughout the life of a system to ensure that all requirements are being met. Assessment and testing are a continual process rather than just a one-time event at some point during the life of a system. There are many areas that should be subjected to some type of evaluation, assessment, or formal testing, but in order to do this, clear, tangible requirements must be placed on the program and the system. The four major areas that have been presented in this book and shown in Figure 13-1 are the supportability engineering process of the buyer and seller, the supportability characteristics of the design, the physical support resources for the system, and the adequacy and responsiveness of the support infrastructure where the system will be placed. Each of these areas has specific things that should and must be assessed.

SUPPORTABILITY ENGINEERING PROCESS

The preceding chapters of this book have described a process that is proven to integrate supportability into the design of a system by participating in the critical decision-making events throughout system development. Each of these supportability engineering activities must be assessed to ensure that all possible benefits can be realized through the process. The overall process, restated in Figure 13-2, shows not only what should be done when but also what should be assessed.

The Buyer Process

The success of supportability engineering rests with the buyer. Therefore, it is extremely important that the buyer continually perform self-assessment of the supportability engineering process when applied to a specific program. Certain strategic points can provide an indication of future success or failure. These strategic points are listed in Figure 13-3. The first and most important assessment point is the definition of the user's need. This definition must contain a measurable description of the mission profile, including mission scenario, mission time, mission events, environment of use, and any other factor that would have a bearing on the utility of the system. Second, the specification issued with the request for proposal (RFP) must contain all the measurable supportability characteristics that have been described in this book, especially those pertaining to reliability, maintainability, testability, and standardization. The issues consolidated in Figure 13-6 are the minimum technical requirements that must be determined and included in the specification. Third, every potential seller's proposal must include a description of how each of the measurable requirements in the specification will be achieved in the proposed design solution. Fourth, every measurable supportability criterion must be included in the contractual requirements between the buyer and seller. These requirements must be of equal importance with any performance criteria. Fifth, every design review, especially preliminary design

- Supportability engineering process
- Supportability characteristics of the design
- Support resources
- Support infrastructure

FIGURE 13-1   Supportability assessment and testing.

review (PDR) and critical design review (CDR), must address the evolving design's ability to meet all supportability requirements. Sixth, the prototype testing must include supportability as a key assessment issue. Finally, supportability must be as important as performance for final acceptance of the system design. Successfully passing each of these assessment points should result in a supportable design. The methods to be used to perform this process assessment and the persons responsible to perform the assessment must be described in the buyer's supportability engineering management plan discussed in Chapter 15.

The Seller Process

The seller also should perform a continuing self-assessment of the application of the supportability engineering process as the design is created. Figure 13-4 shows the strategic assessment points. All supportability characteristics required by the buyer's specification must be included in the proposal submitted in response to the RFP. The proposal should include an estimate or prediction of what the proposed design will achieve and the methods that will be used to make sure that these characteristics actually happen in the final design solution. The seller's systems engineering process should include supportability engineering as a participating member of the team. The supportability requirements should have equal weighting when making any design tradeoff decision. All instructions issued to design engineers and purchasing specialists must include measurable supportability criteria that can be linked back to the total system supportability requirements. Individual goals, thresholds, and constraints must be mandated clearly as requirements for successful acceptance of the design and procurement solutions. Finally, supportability should be included in every test of the system—engineering models, prototypes and the final design. Any and all methods to be used to perform this assessment must be described in the seller's supportability management plan discussed in Chapter 15.

DESIGN SUPPORTABILITY CHARACTERISTICS

Achievement of a supportable design is the responsibility of systems engineering. Supportability engineers are key participants in the systems engineering process to ensure that all supportability characteristics are present in the final design solution. Thus actual assessment of the supportability of a design must be at a higher level than just supportability engineering. Assessment should be both informal and formal.

Informal Assessment

Supportability is not something that is viewed only by supportability engineers. Everyone should participate in ensuring that a design meets its supportability goals, thresholds, and constraints. Therefore, all these goals, thresholds, and constraints should be communicated to all organizations

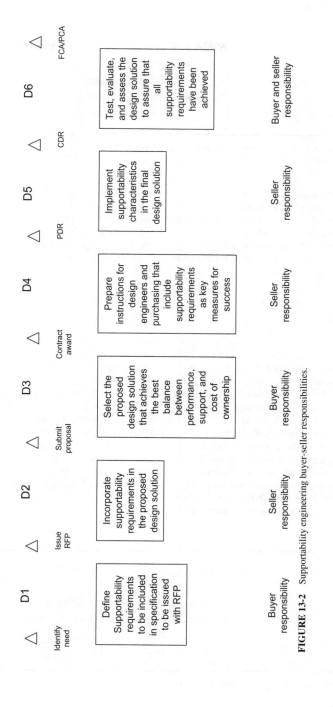

FIGURE 13-2 Supportability engineering buyer-seller responsibilities.

- User need
- Specification
- Proposal
- Contract
- Design reviews
- Prototype testing
- Acceptance testing

FIGURE 13-3 Strategic buyer supportability assessment points.

and persons associated or affiliated with a system design process. Everyone should be tasked to continually view the evolving design solution to make sure that everyone agrees that the final design contains all supportability characteristics. The systems engineering program and the management processes both should have the facility for anyone to provide informal assessment of the supportability of the design and to suggest possible improvements. Every meeting and review, both internally and externally, should have supportability on the agenda as an open forum when any supportability issues can be raised.

Formal Assessment

Formal assessment is always required to ensure that a design meets all supportability requirements. There are four generic methods, as shown at Figure 13-5, that can be used for assessment. An *analysis* is any technical process that compares a requirement with a theoretical estimate or best engineering judgment of a future outcome. An analysis may use expert opinion as the basis for its results. Typically, analyses are performed by the seller as an interim assessment of the potential of the evolving design solution to meet supportability requirements. An *evaluation* is the same as an analysis, with the exception that it is performed by the seller and may be at a later stage of the development cycle or at a higher level than an analysis. A *demonstration* is an event conducted jointly by the buyer and seller to show achievement of some technical aspect. A demonstration normally requires active participation of representatives from both the buyer and the seller organizations. It also normally requires representative examples of the final physical design. *Testing* is the most formal of the assessment methods. It may be performed by either the buyer or the seller but is used to show attainment of measurable supportability criteria. Typically, the test is planned to achieve specific objectives, and a clear pass or fail criteria is established.

- Proposal
- Contract
- Design engineer instructions
- Purchasing specialist instructions
- Subcontracts
- Acceptance testing

FIGURE 13-4 Strategic seller supportability assessment points.

- Analysis
- Evaluation
- Demonstration
- Test

FIGURE 13-5   Assessment methods.

Buyer Assessment and Testing

All buyer-performed assessments must be planned. The initial planning for assessment starts during the first phase of procurement. Any supportability requirement that is contained in the specification must be included in the overall buyer assessment plan. The plan should cover development and operation of the system. All requirements for supportability must be capable of being assessed when the system becomes operational. Therefore, from the buyer's perspective, all testing should focus on the final design solution. There are several names for the buyer's assessment plan. The most common name is the *test and evaluation master plan* (TEMP). An example outline for a TEMP is provided in Figure 13-6. Every aspect of supportability assessment must be included as an integral part of each test contained in the TEMP. An individual test plan must be written for each test listed in the TEMP. In the case of the example plan outline in Figure 13-6, there would be a separate development test plan and a separate operational test plan. Typically, development testing focuses on the system when it is in the factory, laboratory, or manufacturing environment, whereas operational testing moves the system into its actual operating environment. A system should complete all development testing before it starts operational testing. In some cases, development testing is performed by the seller and witnessed by the buyer, and then all operational testing is performed by the buyer with no seller involvement; however, this depends on the specifics of the situation, the program, and the use of the system.

Seller Assessment and Testing

The seller is also responsible for performing whatever analyses, evaluations, and tests are necessary to ensure that the system meets all supportability requirements. There may be some assessment events that the buyer requires the seller to perform, and therefore, the seller should expect payment for these events. Normally, the buyer will require the seller to prepare a master testing plan called an *integrated test plan* (ITP) that identifies and incorporates all assessment activities into the overall engineering and manufacturing of the system. The ITP should coordinate all seller-performed testing. The seller-performed assessment should complement but not duplicate buyer-performed assessment. The individual seller assessment should include all aspects of the design development process. Every event identified in the ITP should incorporate assessment of the supportability characteristics of the design.

Supportability Demonstration

Formal assessment of the achievement of all supportability requirements for a system is normally performed by conducting a supportability demonstration. This should be the last event of qualification testing or acceptance testing of the system. It normally occurs just before the functional configuration audit (FCA) is performed. The demonstration is performed using the qualified system that has completed all performance testing. The steps involved in conducting the demonstration

- Part 1–System introduction
  - a. Mission description.
  - b. System description.
  - c. System threat assessment.
  - d. Measures of effectiveness and suitability.
  - e. Critical technical parameters.
- Part 2–Integrated test program summary
  - a. Integrated test program schedule
    - (1) Major test and evaluation phases and events.
    - (2) Event dates.
    - (3) Decision to proceed dates.
  - b. Management
    - (1) Responsibilities of all T&E participating organizations.
    - (2) T&E structure.
    - (3) Proposed or approved performance exit criteria.
- Part 3–Developmental test and evaluation
  - a. Developmental test and evaluation overview.
  - b. Future developmental test and evaluation.
- Part 4–Operational test and evaluation
  - a. Operational test and evaluation overview
    - (1) The primary purpose of operational test and evaluation.
    - (2) Program schedule, test management structure, and required resources.
  - b. Critical operational issues
    - (1) Critical operational issues.
    - (2) Critical technical parameters and thresholds.
    - (3) Measures of effectiveness and measures of performance.
  - c. Future operational test and evaluation
- Part 5–Test and evaluation resource summary
  - a. Key test and evaluation resources
    - (1) Test articles.
    - (2) Test sites and instrumentation.
    - (3) Test support equipment.
    - (4) Threat representation.
    - (5) Test targets and expendables.
    - (6) Operational force test support.
    - (7) Simulations, models, and testbeds.
    - (8) Special requirements.
    - (9) Test and evaluation funding requirements.
    - (10) Manpower/personnel training.
  - b. Time-phased test and test support resources

FIGURE 13-6 TEMP outline.

- Demonstration plan
- Fault catalog
- Fault stratification
- Fault insertion plan
- Training
- Demo setup
- Conduct demonstration
- Report

FIGURE 13-7   Conducting a supportability demonstration.

are shown in Figure 13-7. Conceptually, the demonstration is conducted by taking a fully functional system, interjecting one failure at a time, and then demonstrating that the system can be supported using the necessary support resources to restore it to an operable condition. The demonstration may take several days to complete. The demonstration should focus on on-system maintenance activities. This is to ensure that the operational availability goal for the system can be achieved and sustained.

Demonstration Plan.   The demonstration plan provides the framework for successful completion of the demonstration. It identifies the responsibilities of both the buyer and the seller to plan and conduct the demonstration. The buyer may require the seller to submit a preliminary demonstration with the proposal to make sure that the approach to showing that the design meets all supportability requirements is reasonable and attainable. Figure 13-8 shows a support demonstration plan outline.

Fault Catalog.   The demonstration focuses on rectifying failures in the system. The failure modes, elements, and criticality analysis (FMECA) discussed in Chapter 5 is the basis for identification of all possible faults that the system may encounter during normal operation. Remember, a *fault* is a physical occurrence that leads to a functional failure. Every failure mode on the FMECA should be reflected in a fault that appears on the fault catalog. This listing of possible faults normally is prepared by the seller.

Fault Stratification.   Fault stratification organizes the faults listed on the fault catalog in order of criticality and frequency of possible occurrence. Stratification uses the results from the reliability prediction to determine which faults may happen most frequently, and the results of the safety analysis to establish the most critical tasks that may create a hazard. Finally, the maintainability prediction is used to identify which maintenance actions may require the largest investment in personnel. All these are combined into the maintenance actions that should be highlighted during the demonstration. Normally, the seller is allowed to select 200 faults from the stratified list, and the buyer adds an additional 100 faults selected from the list, so the result is 300 faults for the demonstration. On major systems the number of faults may be increased to a total of 450 or even 600. Small systems being demonstrated may only require fault lists of 60 to 75 faults. It is up to the buyer to determine the appropriate number of faults to be used during demonstrated.

Fault Insertion Planning.   Again, the demonstration is performed by taking a fully functional system and inserting one fault at a time and showing that the fault was rectified and the system returned to an operable condition in the amount of time specified. This requires that a method be identified for inserting every fault on the stratified fault list into the system. There are various ways of inserting faults. These include using prefaulted modules, rejected assemblies from manufacturing, items that

1. Introduction
2. Demonstration management
   a. Organization
   b. Buyer organization
   c. Seller organization
   d. Participants
      i. Buyer
      ii. Seller
      iii. Subcontractors
      iv. Outside agencies
   e. Responsibilities
3. Demonstration
   a. Supportability characteristics
      (Identification of individual characteristics)
   b. Support resources
      (Identification of each resource, i.e. manuals, tools)
4. Fault catalog development
   a. FMECA
   b. RCM
   c. Other
5. Fault stratification methodology
   a. Data sources
   b. Modeling
6. Fault insertion plan
   a. Prefaulted modules
   b. Failed modules
   c. Simulation
7. Maintenance training
   a. Operator training
   b. Maintainer training
   c. Supervisor training
8. Support resource planning
9. Maintenance manuals
10. Support and test equipment
11. Facilities preparation
12. Demonstration setup
13. Conduct of demonstration
   a. Organization
   b. Fault insertion
   c. Event participation
14. Recording results
15. Scoring
16. Demonstration report
   a. Preparation
   b. Approval
   c. Distribution
17. Schedule
   a. Demonstration preparation
   b. Conduct
   c. Reporting

FIGURE 13-8   Support demonstration plan (outline).

have been damaged during testing, and other ways depending on the technology of the system. Design engineers are the best source for starting to plan how to insert faults for the demonstration. These then need to be refined by systems engineers.

Performing the Demonstration. The demonstration is conducted by the buyer, with maintenance actually performed by user operation and maintenance personnel. Thus these people must be trained prior to the start of the demonstration. They should be given the standard training courses that all future operators and maintainers will receive. Then they are provided with all resources (i.e., tools, manuals, spares, etc.) that have been identified as being required to support the system. Each fault is inserted in the system, and these fully trained people must restore the system within the allotted time. It is common to use a stopwatch to measure that actual time taken for troubleshooting and diagnosing the problem, completing the repair, and then performing a functional test to ensure that the system is again fully functional. It is not necessary to always demonstrate the ability to repair all the faults on the fault list. Performing 300 tasks is a long and expensive process that may take several weeks and require many people. The buyer must make a value decision as to the appropriate number of faults to actually be demonstrated. In cases where the seller successfully demonstrates the first 10% of faults, the buyer may choose to spot check others and then end the demonstration. In cases where the seller can not successfully demonstrate the first 10%, the seller may stop the demonstration and direct system redesign or changes to the support resources. When the seller has completed the redesign or changes, the seller may re-start the demonstration. The process for starting and stopping the demonstration is a contractual issue that must be agreed at the start of the project.

Demonstration Report. The demonstration report is an extremely valuable document that should be used to communicate supportability of the system to all concerned persons and organizations. The report normally is written by the buyer organization and then distributed to anyone associated with system acquisition or operation. The report may be the basis for design improvements or support-resource modification either before or after delivery of the system. At a minimum, every specification requirement pertaining to supportability should be addressed during the demonstration and contained in the final supportability demonstration report.

SUPPORT-RESOURCE ASSESSMENT

The support resources that will be provided for the system in its operational environment also must be assessed for technical accuracy and adequacy. This means that two different issues must be addressed. First, there must be assurance that the physical resources that have been identified as being required to support the system are technically correct. Then there should be a confirmation of the quantity and usability of the resource package that will be delivered with the system.

Maintenance Task Validation

The maintenance task analysis (MTA), described in Chapter 11, should be the sole source for identification and quantification of all support resources for the system. This analysis is typically performed on the "paper design" as it is being created. The only way to check the accuracy of this analysis is to physically perform each maintenance task as physical hardware becomes available. The act of checking is called *maintenance task validation.* Normally, validation is performed on prototype items or preproduction models. In some cases, early engineering models may be available that are sufficiently similar to the final design to allow task validation. The purpose of this validation is to prove the technical accuracy of the MTA that has been recorded in the logistics information system because this information later will become the only source of technical data for preparation of the final deliverable support resources. Task validation normally is performed by the seller and witnessed

---

Is the maintenance task name accurate?
Is the maintenance task name consistent with other similar tasks?
Is the location correct?
Is the task setup correct?
Are the task steps sequentially correct?
Is the task completely described?
Is the task narrative sufficient to prepare maintenance instructions?
Is the task narrative sufficient to prepare training materials?
Is each step time accurate?
Is the total task time accurate?
Is the correct person performing each step?
Is each person's participation identified?
Are all parts and materials required to perform the tasks identified?
Are all tools required to perform the task identified?
Are all items of support and test equipment correctly identified?
Is the total task time correct?
Have all task cautions and warnings been identified?
Are cautions and warnings consistent with similar tasks?
Are all support resources accurately documented?
Have preceding and follow-on tasks been correctly linked?
Can the maintenance task be linked to the FMECA or RCM?
Have all source or reference documents been identified?
Is the illustration for the item being maintained correct?
Has the task been analyzed by safety engineering?
Has the task been analyzed by human factors engineering?

FIGURE 13-9   Maintenance task validation.

by the buyer so that both agree on the technical accuracy of the source for the final resource package. Figure 13-9 shows the major focus of the validation process. Maintenance task validation normally occurs between CDR and first-article testing but can occur earlier on existing subsystems and lower-level items within the system architecture.

Support Resources

The physical support resources that have been developed for the system also must be assessed to ensure that the right resources will be at the right place at the right time in the right quantity to sustain system operation. This includes all resources required to operate and maintain the system. Any resource that has been identified through any method must be included in this assessment. Normally, the assessment is performed by subject matter experts (SME) who have specific qualifications and experience in developing and using the items. Specifically, the range of resources must include all materials, personnel, training, support equipment, and facilities. Other areas that may be included depend on the technology baseline of the system being developed. There must be an agreement between the buyer and the seller as to the range of items to be assessed, and this agreement normally is included in the contract.

Materials

Any materials required to operate or maintain the system must be identified, purchased, and prepositioned so that they are available for use when needed. Operational materials may include fuel, water, food, consumable products, and other items. Maintenance materials typically include spares, repair parts, consumable items, and raw materials. Identification of materials normally is done using the physical supportability analysis process described in Chapter 11. The quantity required of each material should be determined using the appropriate modeling technique that combines demand rate and probability of not running out of the material. Several methods can be used for this purpose. It is always the buyer's responsibility to ensure that adequate quantities of resources are available to support the system.

Support Equipment

Operation or maintenance of a system may require specific items of support equipment. These should be common with other systems to reduce cost of ownership. Every item identified should be clearly justified and should be accepted only if there is no other alternative. The idea of designing a system to avoid requirements for support equipment was presented and discussed in detail in several previous chapters. The requirements for support equipment should be assessed as a part of the supportability demonstration.

Personnel

Personnel requirements tend to be the most expensive support resource required to operate and maintain a system. Requirements for operational people are determined directly by the design of the system, and requirements for maintenance people are determined based on the design of the system, its potential to fail, and the participation of people to restore the system to an operable condition. Assessment of people should address two issues: number of people available and minimum technical capabilities. The number of people required is calculated using the system requirements, the operational concept, and the maintenance concept. These are the basis for numbers of people. The technical qualifications are determine by the complexity of the tasks that the people are required to perform. Each of these areas should be assessed by analysis and evaluation.

Training

The adequacy of training should be assessed continually throughout the final stages of system development. The first assessment is through development of the training needs analysis, which is an output of maintenance task analysis. The goodness of the training course curriculum is assessed by evaluation as it is being constructed. The final assessment for operation and maintenance training is conducted as part of qualification testing for operation training and the supportability demonstration for maintenance training.

Facilities

Assessment of facilities is difficult to perform. Typically, requirements for facilities are based on the operation and maintenance actions that are to be performed at the facility. Thus the assessment looks at the physical adequacy of the facility for task performance. The consolidation of all tasks and their frequency is used to estimate the workload at a facility. However, facilities frequently are shared by several different systems, so capacity sharing becomes a significant issue. In addition, facilities tend to be owned by other organizations, and the user is simply a tenant. This means that facility requirements must be coordinated with organizations that are outside the acquisition of the system.

SUPPORT INFRASTRUCTURE

Assessment of the support infrastructure is difficult. Since the support infrastructure for a system pre-exists the system, assessing its adequacy may be limited to identification and quantification of its limits to support the new system. In most cases, this is an accommodation and acceptance rather than an assessment of improvement. As was discussed in previous chapters, every organization has an existing support infrastructure. It is frequently a limiting factor from system procurement rather than an area of drastic improvement. There may be cases where improvement is possible if the improvement can benefit many systems; however, significant change normally comes with a significant price tag. The money for the change in support infrastructure comes from outside the budget of any single acquisition program. It is only when the overall organizational structure shifts that significant changes can occur to the support infrastructure. Three areas that should be studied as a system nears initial delivery are support processes, support locations, and response times. These three areas are interrelated but can be assessed individually to identify any major issues that should be raised to the appropriate organizational level. The system requirements study discussed in Chapter 3 should be the basis for initiating this assessment.

Support Processes.   The efficiency and responsiveness of the support infrastructure to provide sustainment of the system is critical. The processes that have been established to respond to support demands should be assessed continually to determine their effect on long-term support of the system. The most significant areas that tend to have a big effect on system capability are the processes for the user obtaining operational and maintenance resources and for obtaining assistance from higher or outside agencies. Each of these areas should be assessed continually throughout the procurement program.

Support Locations.   The physical location of support has a bearing on sustainment of the system. Support can be colocated with the system, at a regional site, or at a large consolidated location that may be very distant. The lines of communication and delivery can have a significant effect on the quality and timeliness of support. Having all support colocated with the system normally is not reasonable if there are many systems in use. Conversely, having a single consolidated location may be the most cost-effective alternative to limit the cost of the support infrastructure, but it may be detrimental to timely support. Level-of-repair analysis, discussed in Chapter 11, provides a beneficial method to assess the options for support locations.

Response Times.   The most critical issue when assessing the support infrastructure is determining the adequacy of response times to demands for support. The amount of time spent waiting for materials, fuel, technical assistance, or other support is reflected in the operational availability of the system. The response times experienced by a system normally will be the average time experienced by all systems supported by an infrastructure. For example, if it takes one system five days to receive a part that has been requisitioned, then it will take the same amount of time for all systems. A similar statistic should be assessed for each support commodity required for the system when in operation.

OPERATIONAL AVAILABILITY

The final method of assessing the supportability of a system is to use operational availability. Operational availability is the only management statistic that combines both the design of a system and the support infrastructure into a single measure of goodness. The typical method of calculating A_O allows each element of the design and support to be segregated and at the same time consolidated into a total picture. Every program should start with an A_O target that is developed in the first phase of the life cycle. Every month an estimate should be calculated as to what may be achievable at that point in terms of A_O. The estimate then is compared with the target value. By comparing the estimate with the target, areas of concern should be easily identified for future focused efforts. Continually estimating A_O throughout the acquisition phase of the system life should enable all management and technical decisions to be assessed to determine if the results of the decision will cause A_O to be better

Mission change
Design deficiencies
Support shortfalls

FIGURE 13-10   Operational availability failure.

or worst. This is probably the most beneficial management-level indicator of the progress of the acquisition program in meeting the user's need.

Operational availability provides an invaluable and consistent method to continually assess a system because it allows each piece of the puzzle to be assessed individually as well as the overall situation to be viewed. There are three basic reasons, listed at Figure 13-10, that a system is not meeting its operational availability target. Each of these reasons is an independent issue that can be isolated by operational availability.

The first issue that must be assessed is mission change. Starting in Chapter 1 of this book, the system mission has been the basis for all supportability engineering activities. A change in mission can invalidate every activity if the change significantly effects how the system must perform to be successful. For example, if a person purchases a small car with a small engine solely as transportation to work, but then uses the car to also tow a heavy trailer in the desert or mountains on vacation, this change in use can be detrimental to availability. Mission changes tend to be one of the most significant reasons for operational availability shortfalls.

Designs that are deficient in reliability, maintainability or testability experience more nonavailable time because of longer maintenance periods. Each of these engineering disciplines can be measured, as was shown in Chapter 5 of this book. Accurate in-service data collection allows mean time between failure (MTBF), mean time to repair (MTTR) and fault detection and fault isolation to be measured. Access time can also be measured in-service. Design-to or buy-to targets were developed in the first stage of acquisition for each of these technical characteristics. A significant negative deviation between the targets and actual in-service experience can have a tremendous effect on operational availability. Design deficiencies can be remedied through design changes, modifications or upgrades.

Support shortfalls are the third reason for less than desirable system operational availability. Again, in-service data collection provides the ability to understand which area of the system support infrastructure. Support shortfalls can be difficult to diagnose since the support infrastructure is shared by many systems. There can be several different reasons for shortfalls, but the most typical is lack of sufficient funding. Failure to purchase adequate support resources can cause disruptions in supplying items when needed. There are several other reasons for resource shortfalls, but one that can be very frustrating is where another system is impacting the system being analyzed. For example, system A shares a maintenance technician with system B. System B was designed to have an MTBF of 1000 hours, but actual has an in-service MTBF of only 100 hours. This means that the technician will be required to provide 10 times as many hours of maintenance to system B as was estimated. So, the technician will not be capable of providing the same level of support to system A because his time is being used for system B. A similar analogy can be made where two or more system share a common pool of spares.

Operational availability analysis provides the ability to understand the real reason for system problems. For example, a system that is not available due to the lack of a spare required to repair the system would traditionally trigger a process to increase spares quantities because that is the visible reason for availability failure. However, the lack of a required spare may be the symptom of the problem, but not the true cause of the problem. Investigation of this situation must start with the basic question, "is the system being used within the conditions for which it was designed" If the answer is yes, then the problem is either design or support, but if the answer is no, then it is the mission changes that has caused the shortage of required spares. In this case the only way that the problem can be resolved and avoided in the future is to increase the overall spares quantities held to support the system. However, it should be noted so that when the current system is replaced, this fact can be highlighted as a cause for lower than anticipated operational availability, otherwise the problem will be repeated for the replacing system.

Operational availability is an invaluable measure of system supportability. It should be assessed during every phase of the system life cycle, and it should be used as one of the most important management indicators throughout the operational life of a system.

Supportability requirements compliance matrix						
Supportability requirement	Compliance method				Pass/fail	
	A	E	D	T	P	F

Note – A – Analysis
     E – Evaluation
     D – Demonstration
     T – Test
     P – Pass
     F – Fail

FIGURE 13-11   Supportability requirement compliance matrix.

RECORDING THE RESULTS

The assessment and testing of supportability must always be documented to provide a audit trail of all requirements from identification to completion. The most common method of documenting these results is using a requirements compliance matrix. An example of such a record is provided at Figure 13-11. Use of this method of recording assessment and testing results provides visible evidence that all requirements have been met or it provides identification of further work to be done in order to eventually meet all requirements. A copy of this matrix should be provided to all buyer, seller and user organizations to coordinate supportability related issues throughout the system life. It is very valuable during any design review, program review or audit.

CHAPTER 14
SUPPORTABILITY IN SERVICE

The initial supportability engineering activities culminate with delivery of the system. After delivery, focus shifts to sustaining the mission capability of the system. There are several areas where supportability engineering should have a significant participation throughout the operational life of the system. These are shown at Figure 14-1. The first area, operation and support information collection, is necessary to perform all the other activities indicated. The basis for all in-service supportability engineering activities is to achieve all the goals, thresholds, and constraints developed during acquisition. Every supportability characteristic developed and used during product design should be measurable in service. It is extremely difficult to receive the full benefit from in-service supportability engineering without this as the basis for all activities.

INFORMATION COLLECTION

All decisions made during acquisition of the system were based on goals, thresholds, and constraints that were developed through comparison with similar systems, historical data on previous systems, and predictions of possible future results. Now that the system has entered service, it is necessary to collect information about what really happens. All information collected must be accurate, timely, represented in a way that is conducive to analysis, and most important, believable. Every organization has its own preferred method of in-service information collection; however, at a minimum, any information-collection system should address the points listed at Figure 14-2. There should be an appropriate place to capture all the information collected. ISO 10303, *Standard for Exchange of Product Model Data (STEP),* contains Application Protocol 239, "Product Life Cycle Support (PLCS)," which is a comprehensive information repository. PLCS has the capability of holding all data developed during acquisition and to record information that is developed during service, as shown at Figure 14-3. GEIA Standard 927, *Common Data Schema for Complex Systems,* is the data protocol for PLCS and GEIA Standard 0007, Logistics Product Data, is the data standard for implementation of PLCS. Supportability engineering can use PLCS during in-service as the primary source for information to subject to the activities described in the remainder of this chapter.

MISSION ANALYSIS

Systems are acquired to meet a specific need. The need, which normally is expressed as a mission statement, is the starting point for all supportability engineering activities that have been described in this book. One of the major reasons that systems fail to meet all expectations in service is that the mission changes over the operational life of the system. In some cases, owing to the length of time between the start of acquisition and system delivery, the mission has changed by the first day of service. When the mission changes, all decisions that were made during acquisition may be affected. Changes in use profile can have serious effects on system life and support requirements. Also, changes in the environment of use may create significant problems for the system and for support. Supportability engineering continually must compare the

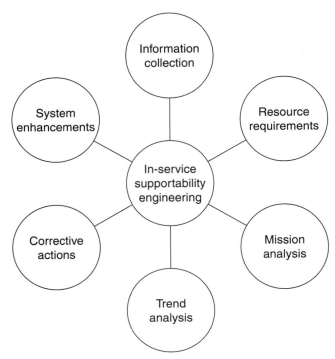

FIGURE 14-1 In-service supportabiity engineering.

- Event being recorded (operation or support)
- Location event occurred
- Environmental conditions
- Time and date of event
- Description of event
- Circumstances surrounding the event
- Recurring or nonrecurring event

FIGURE 14-2 In-service information collection.

- Support engineering—provide and sustain the support infrastructure

- Resource management—plan and record resource utilization against task

- Configuration management—manage change throughout the product life cycle, with the provision for tracking product configuration information

- Maintenance and feedback—maintain, test, diagnose, calibrate, repair, modify physical product, including schedules, resources, and feedback

---

FIGURE 14-3 PLCS information collection and use.

- Change in mission
  - Different stresses
  - Different failure modes
  - Loss of functionality
  - Lower effectiveness
- Increased usage rate
  - Lower reliability
  - Lower availability
  - Shorter system life
- Change in environment of use
  - Higher failure rate
  - Not designed for unique conditions
  - Incapable of mission completion

FIGURE 14-4   Mission analysis focus.

current mission profile, usage rate, and environment of use with those established as the procurement base-line at the start of acquisition. Figure 14-4 provides examples of these issues. Any deviation from the base-line should signal a possible potential for system shortfalls and support problems. These may result in lower operational effectiveness and availability while increasing cost of ownership. This should be one of the major concerns of supportability engineering throughout the in-service life of any system.

RESOURCE REQUIREMENTS

The support resources that were developed during acquisition are based solely on predictions of future requirements based on past history. In some cases where no historical data exist, support resources are developed using only best engineering judgment. This results in support requirements being either well supported or undersupported or support resources being underutilized. The three areas that tend to be most susceptible to these problems are personnel, facilities, and materials. Figure 14-5 shows examples

Resource	Concern	Problem indicators
Personnel	Correct number Adequate skill set	Waiting time Induced error rate
Facilities	Correct capacity Proper capability	Waiting time Moving workload
Materials	Correct items Correct quantities	Not authorized to stock Waiting time

FIGURE 14-5   Resource requirements.

of how resources should be measured. These are the resources that are most expensive to purchase and hold as inventory. Thus each area must be assessed continually to make sure that the proper numbers are available without a significant excess being held. When issues are identified, supportability engineering should be responsible for investigation, analysis, and development of any corrective actions.

TREND ANALYSIS

Some supportability problems take time to become evident. This creates some interesting challenges for supportability engineering. The challenge comes from being able to identify a problem and then to determine the root cause so that the problem can be resolved. Typically, most problems eventually result in decreases in operational availability and increases in cost of ownership. However, waiting until the problem has become large enough to have a significant impact on operational availability or cost of ownership means that the user will have been penalized by having a system that does not meet requirements. Thus it is desirable to have management and evaluation points that can indicate potential problems before they become catastrophic. This means that the information collected on a system must be organized in such a way that it enhances the ability to detect potential problems early. Certain issues hamper trend analysis, including cyclic system use, seasonal fluctuations in environmental influences, budgeting limitations, and simply normal changes in system performance over its useful life. Mechanisms such as reliability bathtub curves, obsolescence curves, and wearout rates allow many of these issues to be accommodated. However, basic trend analysis should provide visibility for each individual factor that may be the cause of problems.

Figure 14-6 shows the concept of establishing a trend analysis process. There always must be a known starting point. This would be the goals, thresholds, and constraints established at the start of acquisition. Some of these points would be expected to increase over time, some would decrease, and

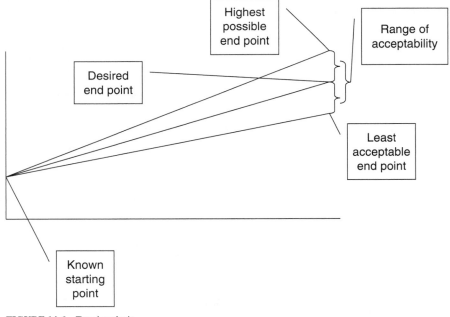

FIGURE 14-6   Trend analysis.

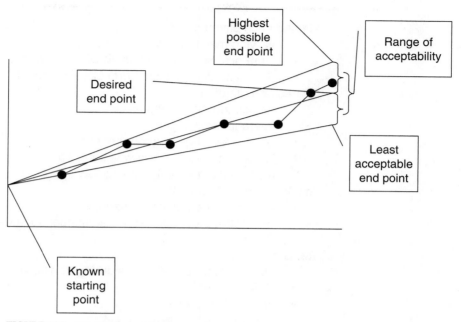

FIGURE 14-7   Trend analysis plot.

some would be expected to remain constant. A trend analysis matrix should be created for each goal, threshold, and constraint. Then the information collected during in-service life should be presented in a manner that is consistent with the point being studied. Figure 14-7 shows how periodic plotting of these results can be put into context with their overall importance and focus efforts in the areas that have the most significant problems. Anytime that the information plotted indicates that the system is nearing the upper or lower edge of the range of acceptability, a focused study should be performed to determine the cause of the deviation. This study always should allow correction prior to the point crossing the upper or lower limits of the range of acceptability.

CORRECTIVE ACTIONS

Supportability engineering should be key participants in all corrective actions taken during the in-service life of a system. Corrective actions can include design changes, changes or adjustments of support resources, and modifications to the support infrastructure. Each of these corrective actions actually goes through a different process for completion and implementation.

Design

All changes to the design after delivery must be made using the change control process described in Chapter 12. No other method of implementing a change to the design should be allowed. Remember that one of the major reasons behind strict configuration control is supportability. Supportability engineering is responsible for analyzing all proposed design changes to determine the effect of the change

Maintenance planning
- Varying maintenance concepts
- More difficult with more levels of maintenance
- Contractor logistics support flexibility benefit

Spares and materials
- Adjustment of sparing levels
- Interchangeability of spares
- Spares leveling by number of systems
- Transition from old to new version systems required management of both old and new spares

Technical documentation
- Change pages to paper manuals
- Interactive electronic technical manual updates
- Serialization effectivity

Training
- Dual training requirements during transition
- Safety focus
- Hazard limitations
- Operational capability

Support and test equipment
- New support equipment
- Test program set modification
- Test equipment compatibility

Facilities
- Operations
- Supply
- Maintenance
- Training
- Interfaces

FIGURE 14-8   Analysis of design changes.

on operational availability and support resources. Figure 14-8 highlights the areas that must be analyzed before any change is approved. A cost-benefit analysis must be performed to determine if the value of the change to the system is greater than the cost of developing and incorporating the design change plus changing the support for the system. It may take a combination of life-cycle cost, through-life cost, and whole-life cost to answer this question.

Support Resources

Changes may be required to the support-resource package that was delivered with the system. As stated previously, the range and quantity of support resources probably was developed using estimates and predictions. These techniques are not perfect, and the final resource package always has areas where improvement is needed. Most often the improvement is in adjusting quantities and locations of materials required for operation and support. This area is also the one that tends to be simpler to achieve. Other areas, such as realignment or reassignment of personnel, may take several years, especially if more people are required. Facility issues never may be resolved completely without a significant upper-level management initiative to push the issue because facility modifications are costly and probably require construction of new or modification of existing buildings.

Support Infrastructure

Modification to the support infrastructure can only be justified when the change will benefit the majority of systems that share the infrastructure. The only proven way to justify such a monumental change is to demonstrate an improvement in operational availability and cost of ownership. An example of such a change is the move of the industry to change from stockpiling materials (just in case they were needed) to the concept of just-in-time. This change was a very significant event. People who were comfortable with the old method of retaining large quantities of materials at many locations had to be convinced that they could trust just-in-time materials management. Implementing a switch normally requires complete renovation of existing policies and procedures, which then is followed by extensive migration from the old business process to the new. This may take years to accomplish. Changes and updates to support infrastructures are a necessary form of progress; however, any change comes with inherent growing problems that initially may make the new process more trouble than the old. It is only when all the problems are resolved that true benefit starts to be received from the change.

SYSTEM ENHANCEMENT

Supportability engineering activities during the in-service life of a system are not limited to just solving problems. They also look for ways to enhance the system, support resources, and support infrastructure. One of the most common ways to do this is to continually research and evaluate new and emerging technologies that may be applicable to the system. This activity was discussed in previous chapters of this book. The discussion presented in Chapter 2 about the system ownership life cycle indicates that a system may be upgraded or modernized at some point during the in-service phase to extend its life and provide greater mission capability through expanded functionality. When an upgrade or modernization is planned and implemented, the system must loop back into the design and development processes that originally created it. Every analysis and every prediction performed for the original development of the system must be revisited to determine what positive and negative issues may arise from the changes. This is the continuing evolving life of the system. There should be a constant search for ways that the system and its required support resources can be enhanced over its useful life.

THE NEXT SYSTEM

The supportability engineering process never stops. The in-service phase of one system leads to the start of planning and development of the next generation system. This requires an organized approach to planning how the in-service activities required to support a system will form the basis for learning how to improve the supportability characteristics of the next system and streamlining or enhancing the support infrastructure to be used in the future. Figure 14-9 lists the significant areas of interest for setting the foundation for the next system.

System Requirements Study
Standardization
Obsolescence and Technology
Support options
Information
Mission analysis
Costs
Lessons

FIGURE 14-9   Foundation for the Next System.

System Requirements Study

The current system is the basis for planning the next system. Information on the current system will be documented in section 3 of the system requirements study for the next system. It is important that every effort be made to accurately document all supportability issues that arise during the in-service phase of the current system. Areas that should be highlighted include; 1. design issues not recognized during development, 2, problems with providing maintenance support; 3, problems with deployability, mobility or transportability; and 4, environmental issues that were not forseen during development.

Standardization

One of the major efforts during system development is maximizing standardization on support resources which should lower cost of ownership and lower the incidence of obsolescence issues. The results of standardization efforts must be assessed to determine successes and failures. The real knowledge of the success of standardization efforts will allow a better result on the next system. A comprehensive list of positive and negative results of standardization should be documented in the program lessons. They should address materials, support equipment, facilities, personnel and other significant resources.

Obsolescence and Technology

Resource obsolescence is always a risk that can no be controlled by the developer or user of a system. However, it can be the most detrimental issues to long-term support and cost of system ownership. Obsolescence is caused by technology evolution. It also occurs when producers of a product close their production line and most users switch to another technology. There is no way that a program can predict when an item or its support resources become obsolete. Figure 14-10 shows that obsolescence can be caused when the technology of other related systems change or when the support infrastructure ceases to provide required materials. Diminishing Manufacturing Sources and Material Shortages (DMSMS) is a formalized concept that focuses on possible obsolescence issues for a system and its long term support. Either case is significantly detrimental to long term support of the system and will assuredly cause its cost of ownership to accelerate out of control.

Support Options

The maintenance and support philosophy of an organization evolves as the systems it owns changes and the technology baseline of the world moves forward. Each successive system acquired must consider this evolving support solution. Systems may have an operational life of 20–30 years which means that the support for a specific system may change over its life. This can be seen when viewing the life of a system in total and comparing it to contemporary systems owned by that organization and similar organizations that may be using the same technology baseline for their systems.

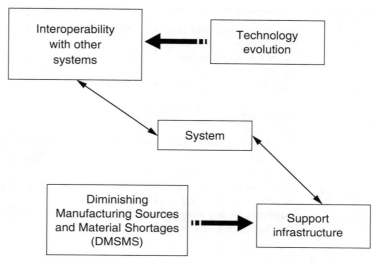

FIGURE 14-10   Obsolescence causes.

Information

The supportability engineering process depends on accurate and timely information on previous systems, mission requirements, and current and future support capabilities. The baseline comparison analysis described in Chapter 11 can only be beneficial if the information analyzed is accurate, realistic and believable. The lack of proper information significantly limits the utility of this technique. Every effort must be made to record every piece of information that can be obtained from system operation and support in order to improve every possible aspect on the next system.

Mission Analysis

Chapter 1 of this book highlighted the importance of understanding the system mission and how the mission affects all aspects of supportability engineering. Typically, missions change over time as the needs of the user change. This can be accelerated by identification of new or evolving requirements that necessitate changes in technology. It is many times possible to project future mission changes so that the changes can be accommodated with as little difficulty as possible. For example, when a family expects a new baby, they can project a requirement for more transportation, more money for education and so on simply because meeting the needs of a child includes all those aspects. Many of these needs require early planning to that when the need occurs the family has the capability to respond without undue hardships on their budget or other resources such as time.

Costs

The estimation of cost of ownership is an integral part of decision-making by supportability engineering. It is important that the real costs incurred be compared with the costs that were estimated. This provides another valuable input into the baseline comparison analysis for the next system. The operational support of a system must be accurately documented in terms of financial expenditures so that the accuracy of cost estimation techniques can be assessed to determine where they can be improved to be better on the next system. Cost estimation will always be subject to the accuracy of any cost basis used for comparison, but the more accurate the comparison figures, the more usable any resulting projects for the next system.

Description of problem (describe in detail the in-service result that has been identified as a problem

Problem measurement (provide any possible method used for measurement of the problem

Goal/threshold/constraint (state the goal/threshold/constraint developed during acquisition)

Deviation (describe the amount of deviation from the goal/threshold/constraint)

Causative factors (describe what factors caused the deviation)

Process factors (describe any supportability engineering process that was used during acquisition)

Decision points (describe the schedule points during acquisition where decisions precipitated the deviation

Recommended alternatives (describe how the deviation may possibly be avoided on the next system)

FIGURE 14-11   Lessons learned format.

Lessons

Lessons learn form an important part of an organization's historical knowledge base that is used for the next system. Development of appropriate and usable lessons learned is a critical part of the overall business process of any organization. It allows improvement and betterment of systems and support infrastructures. However, organizations tend to be very weak in documenting lessons for use by others. This is partly due to the human nature trait of being hesitant to declare substandard personal performance that caused problems. Historically, most organizations document symptoms or problems rather than the cause of the problem. For example, a student in a mathematics course works a math problem and produces an incorrect answer. The fact that the student produced the wrong answer indicates that the student does not know how to work the mathematical function that produced the wrong answer. If the student is not taught the proper way to perform the mathematical function, the student will continually get the answer wrong. So, the teacher must show the proper method to perform the function. Organizations must also be capable of performing supportability engineering functions. Simply stating that support of a system was inadequate is insufficient to correct the problem on the next system. The contents for a lessons learned document should be comprehensive and descriptive of what has actually been learned. The minimum issues for each lesson learned is provided in Figure 14-11.

CONCLUSION

Supportability engineering is a continuous process that enables each successive system to improve operational effectiveness, operational availability and cost of ownership. It is this continuing benefit that validates its application on all systems throughout their entire life.

CHAPTER 15
SUPPORTABILITY ENGINEERING MANAGEMENT

Supportability engineering is a key participant on the systems engineering team on any program. This means that managing supportability engineering consists of effective integration of activities with all others being performed at the same time. It also means that getting results from supportability engineering depends on the coordination and negotiation of supportability issues with the issues from every other engineering discipline participating in the program. Figure 15-1 illustrates this complex series of issues. There is no best solution, only the solution that gives the most reasonable balance among operational effectiveness, operational availability, and cost of ownership. Successful implementation of supportability engineering requires proper interfacing with all other disciplines, performing activities in concert with the program schedule, making sure that the work is completely identified and quantified, and finally, establishing and following well-orchestrated management and implementation plans.

INTERFACES

Typically, the systems engineering team has representation from every discipline required to produce the system. The team membership varies based on the technology of the system being designed but normally has the disciplines listed at Figure 15-2. Each discipline has its own concept as to what the ultimate design characteristics should be to meet the ultimate best results. However, each discipline's ideas are different from those of other members of the team. For example, reliability engineering strives for a design that will not fail, but maintainability engineering assumes that it will fail, so it looks to repair it quickly. While the disciplines of reliability and maintainability are complementary, they also differ greatly in their views of the best design solution. Supportability engineering looks for simplicity and standardization, whereas some of the other disciplines look for complexity and innovation. It is the give and take among these disciplines that creates the final design of the system.

Thus management of the supportability engineering discipline on a program must address external interfaces with all other disciplines as well as its own internal activities. Many of the supportability engineering concepts actually incorporate issues from many other disciplines, so it is very important that supportability engineering management focus on integration of ideas from all rather than just interjection of supportability issues only. This is especially true when considering that supportability in engineering is the discipline that is responsible for incorporation of support infrastructure issues into the design process, whereas all others normally focus only on the design itself. This makes supportability one of the cornerstone disciplines on any program. Figure 15-3 shows that supportability is an equal participant in determining the system design requirements, but this role must be viewed as being philosophically different from the others because of the added scope of looking to the in-service support of the system.

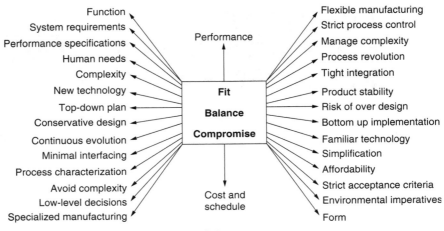

FIGURE 15-1 Differing opinions on the best solution.

- Systems engineer
- Electrical engineer
- Mechanical engineer
- Software engineer
- Human factors engineer
- Safety engineer
- Production engineer
- Supportability engineer
- Reliability engineer
- Maintainability engineer
- Testability engineer
- Quality engineer

FIGURE 15-2 Systems engineering team.

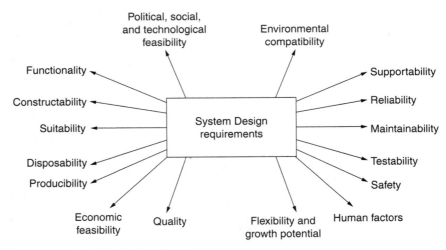

FIGURE 15-3 System design requirements.

SCHEDULE

Timing of supportability engineering activities is crucial for success. This point has been stated and restated throughout this book. The windows of opportunity described int Figure 2-7 for design programs and Figure 2-14 for off-the-shelf (OTS) programs are the keys to success. All supportability engineering efforts must be done in the right window of opportunity or they may simply result in a significant effort in futility. One of the most common reasons for lack of supportability engineering success is attempting to apply the techniques too late for them to have any benefit. Figure 15-4 summarizes the key schedule-related events that depend on supportability success. Each event has

Design program

Issue of specification in RFP
- (Contains measurable requirements)

Proposal submittal
- (Contains approach to meet requirements)

Contract award
- (Requirements become contractual obligation)

Preliminary design review
- (Measurable requirements to all designer, purchasers, and subcontractors)

Critical design review
- (Measurable requirements met in paper design)

Functional configuration audit
- (All measurable requirements achieved)

Physical configuration audit
- (All documentation accurate)

Off-the-shelf program

Issue of specification in RFP
- (Contains measurable requirements)
Proposal submittal
- (Contains evidence that OTS system meets requirements)
Down select
- (OTS system meets technical requirements)
Testing
- (Testing confirms measurable supportability criteria met)
Contract award
- (Selected OTS system meets supportability requirements)

FIGURE 15-4   Key schedule-related events for success.

different supportability activities, and completion of one event leads to the next until the system is delivered. When one or more of the supportability engineering activities is done out of sequence or not done at all, the entire process is jeopardized. This is a very challenging point because most of the supportability engineering activities actually are scheduled and budgeted by systems engineering. This means that systems engineering must be completely schooled on all supportability engineering analysis techniques, understand the benefits of their application, and be aware of when the techniques have the most value in meeting overall system requirements.

SCOPE OF WORK

Supportability engineering is applied as part of the overall system engineering process; therefore, the scope of work for application on any specific program can be determined only by quantifying the participation required to perform the activities described in this book. The majority of supportability engineering efforts can be grouped into a single series of interrelated activities, and these are totally dependent on the level of detail and degree of application. Two different types of supportability engineering activities have been discussed in this book: functional supportability engineering and physical supportability engineering (see Chapter 11). Each of these types of supportability engineering must be quantified differently because they are totally different activities done at different times on a program for different purposes.

Functional Supportability Engineering

Functional supportability engineering activities are performed to interject supportability into the design as it evolves through the different stages of its life cycle. Normally, these are considered level-of-effort activities. This means some number of people working for some period of time to produce some nonquantifiable parts to the overall program. This is the typical approach to determining the scope of any systems engineering work. Therefore, the scope of level-of-effort work again is linked directly to the program schedule. For example, if the amount of time for window of opportunity D1 on a design program is 24 months, then the scope of work probably would be estimated at some number of people participating for 24 months. Figure 15-5 shows how the level of effort is estimated using

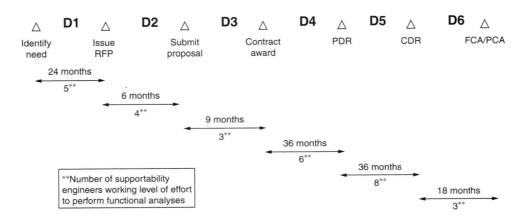

FIGURE 15-5   Functional supportability engineering: scope of work, design program.

Phase	Duration	People	Total
D1	24 months	5	120 months
D2	6 months	4	24 months
D3	9 months	3	27 months
D4	36 months	6	216 months
D5	36 months	8	288 months
D6	18 months	3	54 months
Program total level of effort scope of work			729 months

FIGURE 15-6   Estimating functional supportability engineering level of effort.

the program schedule as the basis for people participating in the program. As long as systems engineering activities are being performed, there must be some level of effort of supportability engineering also being performed. Figure 15-6 shows how these numbers are translated into the total scope of work to perform supportability engineering on a program through all stages of acquisition. (*Author's note*: *The numbers shown in this figure are to illustrate the technique of estimation and should not be used as the actual estimation numbers on a real program.*) It is safe to assume that all functional supportability engineering activities can be estimated as level of effort.

Physical Supportability Engineering

The scope of work to perform physical supportability engineering on a specific program is directly related to the number of maintenance-significant items in the system and the number of maintenance tasks for each of those items. Each organization must develop its own time standard for performing these analyses. Every organization, every product, and every technology tends to have slightly different time standards. Figure 15-7 shows that to make this estimate, six inputs are required, two from system design (number of maintenance-significant items and number of maintenance tasks) and four from the organization (time required to perform maintenance task analysis, time required to validate maintenance task analysis results, time required to perform noneconomic level of repair analysis (LORA) and time required to model system LORA). (*Author's note:* The numbers shown in this figure are to illustrate the technique of estimation and should not be used as the estimation numbers on a real program.) Estimation of physical supportability engineering scope using these six input values is very straightforward. There is no level-of-effort estimation for physical supportability engineering.

PLANNING

Two levels of planning are required for supportability engineering success. First, there needs to be a clear description of all the functional and physical supportability activities that will be performed on a program. This is internal planning to make sure that everything is done in the right sequence to

Input values

Number of MSI in system	250
Average number of maintenance tasks per MSI	3
Time required to perform one MTA	8 hours
Time required to validate one MTA	2 hours
Time required to perform non-economic LORA on one MSI	0.5 hours
Time required to perform system economic LORA	5 days

Calculations

Number of maintenance tasks = number of MSI × average tasks per MSI

$$250 \quad \times \quad 3 \quad = 750 \text{ maintenance tasks}$$

Total MTA hours = (perform time + validate time) x number of tasks

$$(\quad 8 \quad + \quad 2 \quad) \quad \times \quad 750 \quad = 7500 \text{ labor hours}$$

Total noneconomic LORA hours = perform time x number of MSI

$$0.5 \quad \times \quad 750 \quad = \quad 375 \text{ labor hours}$$

Total economic LORA modeling hours = 5 days @ 8 hours =   40 labor hours

Total physical supportability engineering Scope of work   = 8665 labor hours

Author's note—The numbers and times presented by the figure are for illustration purposes only and should not be assumed to be true on a real program.

FIGURE 15-7   Estimating physical supportability engineering scope of work.

provide the best benefit possible. Figure 15-8 provides an outline for a standard supportability engineering implementation plan (SEIP) A SEIP should be prepared for every program. This plan describes all supportability engineering activities required to meet the required outcomes on a specific program. The second level of planning is to integrate supportability with the overall program and organizational requirements. This is normally accomplished using a supportability engineering management plan (SEMP). The SEMP is a higher-level plan and may be oriented more toward the orga-

Supportability Engineering Implementation Plan
(SEIP)

1.0 Introduction.

This section identifies the program, the applicable organization (contractor, government, or both), the date approved, and the revision version. If applicable, the title page should show an internal document control number.

1.1 Plan Executive Summary.

This section describes the technical plan summary and applicability. If applicable, it also lists major subordinate plans.

1.2 Program Summary.

Briefly describe the program, to include complexities and challenges that are addressed by the technical development effort. This section should highlight applicable goals, thresholds and constraints that will or have been developed for supportability.

1.3 Scope.

Describe the applicability of the SEIP and assign specific responsibilities for the required activities shown in the SEIP.

1.4 Management and Organization.

Identify the persons assigned for supportability engineering implementation on the program along with their responsibilities. Describe how supportability engineering fits within the overall systems engineering process. Indicate key relationships for decision making

1.5 Applicable Documents.

List by title, version number, and issue date applicable documents to the technical management efforts described in the SEIP. At a minimum the list should contain the document defining the User Need, product specification, all test plans, acceptance planning and higher management documents.

2.0 Supportability Engineering Process.

This section describes the supportability engineering process to be applied to the program and assigns specific organizational responsibilities for the technical effort, to include contracted or subcontracted technical tasks. This section also details or references technical processes and procedures to be applied to the effort.

FIGURE 15-8   SEIP outline.

2.1 Functional Supportability Engineering Process Planning.

This section identifies the specific functional supportability engineering analyses to be performed. The description of each applicable analysis will include:

1. Name of analysis
2. Personal responsible for performance
3. Goal, threshold or constraint addressed
4. Applicable modeling technique
5. Start and completion dates
6. Inputs required and sources
7. Outputs to be produced
8. Related Supportability engineering activities
9. Related activities of other engineering disciplines
10. Coordination requirements
11. Reporting requirements

The names of operational availability, life cycle cost, through life cost and whole life cost models should be included.

2.2 Physical Supportability Engineering Process Planning

1. MSI definition
2. MTA definition
3. MSI and MTA List
4. Noneconomic LORA criteria
5. Economic LORA model to be used
6. Method and location of data recording
7. Method for accessing support information
8. Integration with design engineering
9. Projected MTA performance schedule
10. Projected MTA validation schedule
11. Integration of Subcontractor MSI/MTA/LORA results
12. Schedule for output to create physical resource package

2.3 Design Interface

This section provides a detailed description of how supportability engineers will interface with the evolving design solution. At a minimum it should contain a description of how supportability engineers will:

1. Participate in the systems engineering process
2. Assure measurable requirements are assigned to each designed, purchased or subcontracted item
3. Assist design engineers and purchasing specialist to achieve their requirements
4. Participate in integration of the supportable design solution
5. Participate in developing operation and maintenance test procedures

FIGURE 15-8   (*Continued*).

2.4 Reviews.

This section addresses supportability engineering participation at all design reviews including PDR and CDR. It should also address supportability engineering participation in design reviews conducted by subcontractors.

2.5 Supportability Demonstration.

This section addresses the methods and tools used to perform the supportability demonstration. At a minimum the following processes and activities should be described:

1. Demonstration plan preparation
2. Fault catalog preparation and review
3. Fault stratification method
4. Fault insertion techniques
5. Responsibility for demonstration setup
6. Participation and responsibilities for conducting the demonstration
7. Demonstration report preparation and distribution
8. Responsibilities for follow-up and corrective actions

2.6 Interoperability Analysis.

This section describes the processes and procedures to be used for formal and informal trade studies, to include supportability and cost-benefit effectiveness analyses which incorporate all external physical and functional interfaces for operation and maintenance of the system.

3.0 Coordination.

This section identifies how supportability engineering activities will be coordinated with all other related disciplines.

3.1 Reporting.

This section should identify all formal and informal methods that will be used to report status and progress of supportability engineering efforts.

3.2 Schedule.

This section should contain a detailed schedule of all supportability engineering activities. The schedule should be linked to other engineering and program schedules.

FIGURE 15-8   (*Continued*).

nization than the program. The SEMP describes how supportability engineering will relate to all other disciplines and management processes. An outline for a typical SEMP is provided in Figure 15-9. Proper planning will ensure that supportability engineering will be applied properly to a program and achieve the benefits required to produce the best balance among operational effectiveness, operational availability, and cost of ownership.

Supportability Engineering Management Plan
(SEMP)

1.0 Introduction.

This section identifies the program, the applicable organization (contractor, government, or both), the date approved, and the revision version. If applicable, the title page should show an internal document control number.

1.1 Plan Executive Summary.

This section describes the technical plan summary and applicability. If applicable, it also lists major subordinate plans.

1.2 Program Summary.

Briefly describe the program, to include complexities and challenges that are addressed by the technical development effort. This section should highlight applicable goals, thresholds and constraints that will or have been developed for supportability.

1.3 Scope.

Describe the applicability of the SEMP and assign general responsibilities for the required activities shown in the SEMP.

1.4 Management and Organization.

Identify the persons assigned for supportability engineering implementation on the program along with their responsibilities. Describe how supportability engineering fits within the overall systems engineering process. Indicate key relationships for decision making

1.5 Applicable Documents.

List by title, version number, and issue date applicable documents to the technical management efforts described in the SEMP. At a minimum the list should contain the document defining the user need, product specification, all test plans, acceptance planning and higher management documents.

2.0 Supportability Engineering Process.

This section describes the supportability engineering process to be applied to the program and assigns specific organizational responsibilities for the technical effort, to include contracted or subcontracted technical tasks. This section also details or references technical processes and procedures to be applied to the effort. Where these processes or procedures are developed as part of the program, the need dates and development schedule should be shown.

FIGURE 15-9 SEMP outline.

2.1   Supportability engineering process planning

This section addresses planning for the key supportability outputs, to include products, processes, and trained people. The following may be included in this section:

— Major products. Include major specification and product baseline development and control

— Design, Purchasing and Subcontracting products. Include a description of how supportability requirements will be incorporated into all design instructions, purchasing instructions and subcontracts

— Supportability engineering process inputs. Include major requirements documents and resolution instructions for conflicting requirements

— Technical work breakdown structure. Describe how and when the technical work breakdown structure will be developed, to include development and tracking tool sets usage

— Subcontracted technical efforts. Describe the integration of contracted and subcontracted technical efforts

— Processes. Describe the use of established technical processes and standards on the program. This should contain at a minimum modeling techniques for operational availability and cost of ownership

— Constraints. List any significant constraint to the technical effort

2.2   Requirements Analysis.

This section describes the methods, procedures, and tools used to analyze program requirements. This section should specify specific tools to be used to capture and trace program requirements.

2.3   Functional analysis/allocation.

This section addresses the methods and tools used to analyze the program requirements and allocate them down into program component functional requirements.

2.4   Synthesis.

This section addresses the methods and tools used to analyze the functional requirements and allocate those requirements to a physical program component.

2.5   System Analysis.

This section describes the processes and procedures to be used for formal and informal trade studies, to include supportability and cost-benefit effectiveness analyses. Also included in this section are the risk management approaches to be used on the program.

FIGURE 15-9   *(Continued).*

2.6   System Control.

This section describes the control strategies needed for the following:

— Configuration management
— Data management
— Interface management
— Schedule tracking and control
— Formal technical reviews
— Informal technical reviews/interchanges
— Subcontractor/supplier control
— Requirements control

3.0   Implementation Planning.

This section outlines the planned activities leading to program implementation. Included in this area may be plans for:

— Supportability test and evaluation
— Transition of technical baselines to operations and maintenance
— Support planning
— Facilities planning
— Operations and user training development
— Program integration into an existing system-of-systems architecture

4.0   Specialty Engineering Planning.

This section outlines all relationships with other engineering discipline. At a minimum it should include reliability, maintainability, testability, safety, and human factors engineering.

5.0   Coordination

This section identifies how supportability engineering activities will be coordinated with all other related disciplines.

5.1   Reporting

This section should identify all formal and informal methods that will be used to report status and progress of supportability engineering efforts.

5.2   Schedule

This section should contain a detailed schedule of all supportability engineering activities. The schedule should be linked to other engineering and program schedules.

FIGURE 15-9   (*Continued*).

MEETINGS AND REVIEWS

Sound management practices require continual communication and coordination among all participants on a program. This is achieved through periodic formal and informal events where ideas are discussed, progress is assessed, and decisions about future activities are reached. Successful programs have a well-orchestrated series of events that provide a forum for all participants to learn of the overall progress of the program and to understand any possible business problems, such as budget and schedule, or technical problems in reaching performance or supportability goals, thresholds, or constraints.

Meetings

Meetings are internal group forums that allow formal and informal coordination. Each discipline may have a weekly status meeting. This would be a very short (less the one hour) informal session to organize daily events. At this level, these meetings normally are technical venues to discuss problems of a technical nature. Typically, weekly group status meetings are held on either Monday or Friday to signify either the start or end of work for the week. Since most activities are planned around the work to be accomplished over a week, this fosters good coordination. Monthly status meetings tend to be more formal and may be chaired by the program manager. These are also internal meetings that tend to address business issues. This type of meeting should never take more than two hours to complete. An important issue to point out is that many organizations have far too many meetings. Meetings can be beneficial but also can be very detrimental if not used properly. Many organizations conduct meeting after meeting on a daily basis as a company culture. As a result, progress stops, and people are being paid to sit in meetings rather than accomplish their assigned tasks. Some organizations waste up to 25 percent of profit by having meetings rather than progressing the program. Any meeting must be justifiable based on the benefit received. Figure 15-10 shows a meeting notification. Specifically

Group: Supportability Engineering
Chair: JV Jones
Location: Conference Room 2A
Date: Monday, 1 April
Time: 0830

Agenda

Subject	Person	Time
Standard tool box contents change	J. Skaines	10 minutes
R/R task for left engine	F. Watson	15 minutes
Progress on demo setup	All	5 minutes

Total meeting time: 30 minutes

FIGURE 15-10   Meeting notification.

identified in this notification are subjects to be discussed, person responsible for leading the discussion, and time allotted for each subject. Note that a total meeting time has been specified. This type of notification is very valuable in managing meetings.

Reviews

Reviews are meeting that involve external participants. Normally, reviews are formal events between two organizations, such as customer and contractor or prime contractor and subcontractor. Reviews can be either technical or business events. Business events normally are called *program reviews* to indicate that all business aspects such as budget, schedule, and contractual issues will be on the agenda. A program review may be held once every three to four months and may be two to three days in duration. These are significant management events. Supportability engineering should be a topic of discussion at every program review. Technical events tend to be more focused on specific issues about the progress of the design efforts. The four most common design reviews held on a program are the system requirements review (SRR), the system design review (SDR), the preliminary design review (PDR), and the critical design review (CDR). Each of these meetings has a specific focus and is held only once during a program.

The SRR is conducted to ensure that all requirements for the system are reasonable, attainable, and understandable. A list of SRR topics is provided at Figure 15-11. The SDR is applicable to very large systems where interface and interoperability are critical to success. This review covers high-level issues that affect the overall performance and supportability of the system. The topics listed at Figure 15-12 show that this review covers all top-level issues for a system design. The PDR is one of the most significant detailed design events on a program. This is where each design engineer or

- Mission and requirements analysis
- Functional flow analysis
- Preliminary requirements allocation
- System/cost-effectiveness analysis
- Trade studies
- Synthesis
- Supportability analysis
- Specialty discipline studies
- System interface studies
- Program risk analysis
- Integrated test planning
- Producibility analysis plans
- Technical performance measurement plan
- Engineering integration
- Functional baseline
- Functional specification
- Agreed design approach
- Targets for improvement
- Thresholds of acceptability

FIGURE 15-11 System requirements review (SRR).

- Functional flow analysis incorporated in top-level approach
- Preliminary requirements allocation expanded
- System/cost-effectiveness analysis
- Trade studies
- Synthesis
- Supportability analysis
- Specialty discipline studies
- Program risk analysis
- Integrated test planning
- Producibility analysis plans
- Technical performance measurement plan
- Engineering integration
- Configuration item identification implementation
- Data management
- System safety
- Life cycle cost analysis
- Preliminary manufacturing plans
- Manpower requirements analysis
- Milestone schedules

FIGURE 15-12   System design review (SDR).

purchasing specialist has received his or her personal goals, thresholds, and constraints for his or her individual piece of the design. The topics listed at Figure 15-13 should be addressed for each item within the system design. The results of the "paper design" process are presented at the CDR. It is at this event that design engineers and purchasing specialists show how their design solutions meet the requirements for the system. Figure 15-14 lists the topics that must be discussed at the CDR to ensure that the design will meet the specification requirements.

SECRETS TO SUCCESS

The ideas discussed in this book have been applied successfully on program after program; however, the success of application can be traced to four very specific things: communication, education, participation, and integration (Figure 15-15). Virtually every group within an organization has been mentioned in this book. This means that internal and external communication of ideas and activities is crucial so that everyone knows what everyone else is doing. Everyone must understand their roles and how they fit with what all others are doing. This requires education on the overall supportability engineering process and the specific areas of implementation.

- Preliminary design synthesis
- Trade studies or design studies
- Functional flow
- Requirements allocation data
- Schematic diagrams
- Equipment layouts
- Environmental control and thermal design aspects
- Power distribution and grounding
- Preliminary mechanical and packaging
- Safety engineering considerations
- Security considerations
- Preliminary lists of materials
- Preliminary reliability and maintainability
- Preliminary weight data
- Interface requirements
- Development test data
- Configuration item development schedule
- Mock-ups or models
- Producibility and manufacturing considerations
- Value engineering considerations
- Transportability and packaging issues
- Human engineering
- Standardization issues
- Existing documentation
- Commercial items or equipments
- Life-cycle cost analysis
- Armament compatibility
- Status of quality assurance program
- Support equipment requirements
- Software functional flow
- Storage allocation data
- Control functions description
- CSCI structure
- Security
- Computer software development facilities
- Development tools

FIGURE 15-13 Preliminary design revirw (PDR).

- Electrical design
- Mechanical design
- Environmental and thermal characteristics
- Electromagnetic compatibility
- Power generation and grounding
- Electrical and mechanical interface compatibility
- Mass properties
- Reliability/maintainability/availability
- System safety engineering
- Security
- Survivability/vulnerability
- Producibility and manufacturing
- Transportability, packaging, and handling
- Human factors engineering
- Standardization
- Design versus support tradeoffs
- Support equipment requirements
- Interface control drawing
- Prototype hardware
- Design analysis and test data
- System allocation document
- Initial manufacturing readiness
- Life-cycle cost analysis
- Quality assurance issues
- Software detailed design document
- Database design document
- Interface design document
- Supporting documentation of analyses, testing, etc.
- System allocation document
- Software programmer's manual
- Firmware support manual
- Progress on CSCI design
- Schedule for remaining milestones
- CSCI testing
- SW quality assurance
- SW configuration control

FIGURE 15-14   Critical design review (CDR).

- Communication
- Education
- Participation
- Integration

Figure 15-15   Secrets of success.

Everyone needs to go to school on supportability. Supportability is not a spectator sport; everyone must participate for the program to be successful. This is where communication and education become so important. Finally, since supportability is a group activity, all efforts from every participating group must be integrated into a single cohesive program. By following these simple ideas, any program can apply the concepts, principles, and techniques that have been presented in this book successfully.

APPENDIX A
SYSTEM REQUIREMENTS STUDY (TYPICAL OUTLINE)

1 GENERAL

This is a stand-alone document that is initiated in windows D1 or O1 of a program. It is distributed to every individual involved with the program, so this section normally provides a description of the program and the basis for consistent understanding of program direction.

1.1 Scope and Purpose

An overall description of the program and organizational strategies

1.2 System Description

Functional or physical depending on program status
Brief but specific
For orientation

1.3 System Mission Profile

Statement of the need
Strategic use
Based on operational requirements

2 QUANTITATIVE SUPPORTABILITY FACTORS

This section should provide focus on the specific areas that are of concern for supportability and set the foundation for all future supportability engineering activities.

2.1 Operating Requirements

Detailed mission/use scenario
Measurement base
Frequency of use
Systems required

2.2 Number of Systems Supported/Fielding Plan

When and where systems will be placed into service
Duration of installation
Setup of support locations
Maintenance
Supply
Training

2.3 Transportation Factors

Deployability/mobility issues

2.4 Maintenance Factors

Significant issues/concerns for providing support

2.5 Environmental Factors

Geographic/climatic issues
"Green" issues (hazardous/toxic)

3 SUMMARY OF SYSTEM BEING REPLACED

This section should contain a detailed description of the system being replaced. The information contained in this section will be the basis for many supportability analyses conducted during acquisition of the new system.

3.1 Operating Requirements

Detailed mission/use scenario
Measurement base
Frequency of use
Systems required

3.2 Number of Systems Supported/Locations

When/where systems operated and were supported

3.3 Transportation Factors

Deployability/mobility problems/solutions

3.4 Maintenance Factors

Significant problems/solutions for providing support

3.5   Environmental Factors

Geographic/climatic problems/solutions
"Green" problems/solutions (hazardous/toxic)

4   SUPPORT TO BE AVAILABLE FOR NEW SYSTEM

This section describes the infrastructure that will be available to support the new system on day 1 of operation. This section does not identify the resources required to support the new system. The information in this section must reflect the support and maintenance philosophy of the user. The information in this section should be inputs from applicable technical disciplines. The information in this section should be very similar or identical to that of other programs going into service at the same time.

4.1   Maintenance Capabilities

Describe how the organization will be performing maintenance for all systems
Should contain the organization's maintenance philosophy
Should contain the program's maintenance concept

4.2   Supply Support

Supply chain management
Wholesale supply organization
Retail supply organization
Materials procurement process
Support to maintenance process

4.3   Personnel

Personnel available
Operation
Maintenance
Supply
Training/training concept

4.4   Facilities

Capacity/utilization rate available
Operation
Maintenance
Supply
Training
Other

4.5   Support Equipment

Standard tool set
Standard tool kit
General-purpose equipment
Operation equipment

Maintenance equipment
Handling equipment

4.6 Test Equipment

List all test, diagnostic, and measurement equipment

4.7 Technical Data

Organizational philosophy for acquisition of data
Operation data
Maintenance data

5 OTHER AVAILABLE SUPPORTABILITY INFORMATION

Any other significant issues or information pertaining to the specifics of the program or situation

APPENDIX B
ABBREVIATIONS AND ACRONYMS

A_A	achieved availability
ACAT	acquisition category
ADM	advanced development model
ADT	administrative delay time
AECMA	association of European constructors or materiel of aerospace
A_I	inherent availability
ALARP	as low as reasonably practical
ALDT	administrative and logistics delay time
A_M	mission availability
A_O	operational availability
AOA	analysis of alternatives
AOG	aircraft on ground
A_S	spares (or supply) availability
ASD	aerospace and defense
AMS	acquisition management system
APO	acquisition program office
ATE	automatic test equipment
ATP	application test program
BAFO	best and final offer
BCS	baseline comparison system
BII	basic-issue item
BIT	built-in test
BITE	built-in test equipment
BOA	basic ordering agreement
BOM	bill of material
CA	criticality analysis
CAD	computer-aided design
CADMID	concept, assessment, demonstration, manufacture, in-service, disposal
CADMIT	concept, assessment, demonstration, migration, in-service, termination

CAE	computer-aided engineering
CAGE	commercial and government entity
CALS	continuous acquisition and life-cycle support
CAM	computer-aided manufacturing
CASREP	casualty report
CBIL	common bulk items list
CCB	configuration control board
CDD	capability development document
CDF	cumulative distribution function
CDR	critical design review
CDRL	contract data requirements list
CFE	contractor-furnished equipment
CFR	constant failure rate
CIL	candidate item list
CLS	contractor logistics support
CM	configuration management
CM	corrective maintenance
CONDO	contractors on deployed operations
CONOPS	concept of operations
C_O	cost of ownership
COO	cost of ownership
COSHH	control of substances hazardous to health
CPD	capability production document
CPFF	cost plus fixed fee
CPIF	cost plus incentive fee
CPM	critical-path method
CRETE	common range electrical test equipment
CRISD	computer resources integrated support document
CSC	computer software component
CSCI	computer software configuration item
CSD	computer software documentation
CSI	contractor source inspection
D	depot
DCN	design change notice
Def Stan	defence standard (UK)
DEMVAL	demonstration validation
DFR	decreasing failure rate
DID	data-item description
DLA	defense logistics agency
DLO	defense logistics organisation
DM	data management
DM	data module

DMRL	data module requirements list
DMWR	depot maintenance work requirements
DPA	defense procurement agency
DRACAS	data recording, analysis, and corrective action system
EAC	estimate at completion
ECP	engineering change proposal
EDM	engineering development model
EMD	engineering and manufacturing development
EOQ	economic order quantity
ESML	expendable supplies and materials list
ESS	environmental stress screening
ETC	estimate to complete
ETM	electronic technical manual
ETP	electronic technical publication
FBD	functional block diagram
FCA	functional configuration audit
FFIP	firm fixed-incentive price
FFP	firm fixed price
FMEA	failure modes and effects analysis
FMECA	failure modes, effects, and criticality analysis
FOC	First of class
FQR	formal qualification review
FRACAS	failure reporting, analysis, and corrective action system
FRB	Failure review board
FSCM	federal supply code for manufacturer
FSD	full-scale development
FSED	full-scale engineering development
FTA	fault-tree analysis
FTE	factory test equipment
G&A	general and administrative expenses
GFE	government-furnished equipment
GFP	government-furnished property
GIDEP	government/industry data-exchange program
GPETE	general-purpose electronic test equipment
GPTE	general-purpose test equipment
GSA	General Services Administration
GSI	government source inspection
HAZOP	hazard and operability analysis
HWCI	hardware configuration item
I	intermediate
IAB	investment approvals board
ICD	installation control drawing

ICD	interface control drawing
ICD	initial capabilities doument
ID	interface device
IETM	interactive electronic technical manual
IETP	interactive electronic technical publication
IFR	increasing failure rate
ILS	integrated logistics support
ILS	intermittent lip service
ILSMRT	ILS management review team
ILSMT	ILS management team
ILSP	ILS plan
IOC	initial operating capability
IP	initial provisioning
IPB	illustrated parts breakdown
IPC	illustrated parts catalogue
IPL	illustrated parts list
IPR	intellectual property rights
IPT	integrated project team
ISD	initial service date
ISIL	interim support items list
ISO	International Organization for Standardization
ISP	integrated support plan
ISRD	in-service reliability demonstration
ISSP	integrated supply support processes
ISSPP	integrated system safety program plan
ITN	invitation to negotiate
ITP	integrated test plan
ITT	invitation to tender
KPP	key performance parameters
LCC	life-cycle cost
LCN	logistics support analysis control number
LD	logistics demonstration
LLCSC	lower-level computer software component
LLTIL	long-lead-time items list
LMI	logistics management information
LO	lubrication order
LOGSA	logistics support agency
LORA	level-of-repair analysis
LRIP	low rate initial production
LRU	Line replaceable unit
LRU	line-repairable units
LSA	logistics support analysis

LSACN	LSA control number
LSAP	LSA plan
LSAR	LSA record
LSD	logistic support date
MAC	maintenance allocation chart
MAC	multiactivity contract
MCRL	master cross-reference list
MCT	mission-capable time
MHE	materials handling equipment
MIL-HDBK	military handbook
MIL-SPEC	military specification
MILSTAMP	military standard transportation and movement procedure
MIL-STD	military standard
MILSTRIP	military standard requisition and issue procedure
MMH/MA	mean man-hours per maintenance action
MMH/OH	mean man-hours per operating hour
MOS	military occupational specialty
MPTA	manpower, personnel, and training analysis
MRSA	material readiness support activity
MSG	maintenance study group
MTA	maintenance task analysis
MTBCF	mean time between critical failures
MTBF	mean time between failures
MTBMA	mean time between maintenance actions
MTBMF	mean time between mission failures
MTBR	mean time between removal
MTBUR	mean time between unscheduled removal
MTTF	mean time to failure
MTTR	mean time to repair
NAMSA	NATO Maintenance and Supply Agency
NATO	North Atlantic Treaty Organization
NETT	new equipment training team
NHA	next-higher assembly
NICP	national inventory control point
NIIN	national item inventory number
NMCT	non-mission-capable time
NRLA	network repair-level analysis
NSN	national (or NATO) stock number
NTE	not to exceed
O	organization
O&S	operation and support
ODC	other direct costs

OJT	on-the-job training
OMP	obsolescence management program
ORLA	optimum repair-level analysis
OST	order ship time
OT	operating time
PAM	preassessment meeting
PBC	performance-based capability
PBL	performance-based logistics
PCA	physical configuration audit
PDES	product data exchange using STEP
PDF	probability distribution function
PDM	preliminary development model
PDM	product data management
PDR	preliminary design review
PERT	program evaluation and review technique
PFI	private finance initiative
PHA	preliminary hazards analysis
PHS&T	packaging, handling, storage, and transportability
PLCS	product life-cycle support
PLT	production lead time
PM	preventive maintenance
PMCS	preventative maintenance checks and services
PMO	program management office
PPB	provisioning parts breakdown
PPDR	preproduction design review
PPDS	preservation and packaging data sheet
PPB	provisioning parts breakdown
PPL	provisioning parts list
PPP	personnel performance profile
PPP	public-private partnership
PPS	provisioning performance schedule
PPSL	program parts selection list
PRAT	production reliability acceptance test
PRS	provisioning requirements statement
PTD	provisioning technical documentation
QA	quality assurance
QC	quality control
QPEI	quantity per end item
QQPRI	qualitative and quantitative personnel requirements report
QVL	qualified vendor list
R&D	research and development
R&M	reliability and maintainability

RAM	reliability, availability, and maintainability
RAMT	reliability, availability, maintainability, and testability
RBD	reliability block diagram
RCM	reliability-centered maintenance
RD/GT	reliability development/growth testing
RFI	request for information
RFP	request for proposal
RFQ	request for quotation
RIL	repairable items list
RLA	repair-level analysis
ROP	reorder point
RPN	risk priority number
RPSTL	repair parts and special tools list
RQT	reliability qualification test
RWO	repair work order
SCD	source control drawing
SCD	specification control drawing
SDL	software development library
SDR	system design review
SDRL	subcontract data requirements list
SE	support equipment
SERD	support equipment recommendations data
SMR	source, maintenance, and recoverability
SOW	statement of work
SPETE	special purpose electronic test equipment
SPS	software procurement specification
SPTD	supplementary provisioning technical documentation
SPTE	special-purpose test equipment
SRD	system requirements document
SRR	system requirements review
SRU	shop repairable unit
SSC	skill specialty code
SSE	support solutions envelope
SSP	system safety program
SSPP	system safety program plan
SSR	software specification review
SSWG	system safety working group
STE	special test equipment
STEP	standard for the exchange of product model data
STTE	special to type test equipment
T&M	time and material
TCM	total corrective maintenance

TDP	technical data package
TDS	technology development strategy
TEMP	test and evaluation management plan
TLC	through-life cost
TLCSC	top-level computer software component
TLIP	through-life information plan
TLMP	through-life management plan
TM	technical manual
TMDE	test, measurement, and diagnostic equipment
TO	technical order
TP	test program
TPI	test program instruction
TPM	total preventive maintenance
TPS	test program set
TRD	test requirements document
TRR	test readiness review
TTEL	tools and test equipment list
URD	user requirements document
UUT	unit under test
WBS	work breakdown structure
WLC	whole-life costs
WRA	weapon replaceable assembly

APPENDIX C
REFERENCE LIBRARY

The following texts are provided as sources for additional information and research concerning supportability engineering and management and related topics.

SUPPORTABILITY ENGINEERING, LOGISTICS ENGINEERING, AND INTEGRATED LOGISTIC SUPPORT (ILS)

Primary References

1. Blanchard, B. S., *Logistics Engineering and Management*, 5th ed. Englewood Cliffs, NJ: Prentice-Hall, 1998 (ISBN 0139053166).
2. Jones, J. V., *Integrated Logistics Support Handbook*, 3rd ed. New York: McGraw-Hill, 2006 (ISBN 0071471685).
3. Jones, J. V., *Logistics Support Analysis Handbook*. New York: Tab Books, 1989 (ISBN 0830633510).
4. Langford, J. W., *Logistics: Principles and Applications*. New York: McGraw-Hill, 1994 (ISBN 007036415X).
5. Orsburn, D. K., *Spares Management Handbook*. New York: Tab Books, 1991 (ISBN 0830676260).

Additional References

1. Ackerman, K., *Practical Handbook of Warehousing*, 4th ed. London: Chapman & Hall, 1997 (ISBN 0412125110).
2. Ballou, R. H., *Business Logistics Management: Planning, Organizing, and Controlling the Supply Chain*, 4th ed. Englewood Cliffs, NJ: Prentice-Hall, 1998 (ISBN 0137956592).
3. Bowersox, D. J., and D. J. Closs, *Logistical Management: The Integrated Supply Chain Process*. New York: McGraw-Hill, 1996 (ISBN 0070068836).
4. Christopher, M., and H. Peck, *Marketing Logistics*. Boston: Butterworth-Heinemann, 1997 (ISBN 0750622091).
5. Copacino, W. C., *Supply Chain Management: The Basics and Beyond*. St. Lucie Press/Apics Series on Resource Management, 1997.
6. Coyle, J. J., and E. J. Bardi, *Transportation*. South-Western Publishers, 1998 (ISBN 0538881801).
7. Coyle, J. J., E. J. Bardi, and C. J. Langley, *The Management of Business Logistics*, 6th ed. West/Wadsworth, 1996 (ISBN 0314065075).
8. Coyle, J. J., E. J. Bardi, and R. A. Novack, *Transportation*, 4th ed. West/Wadsworth 1994 (ISBN 0314028536).
9. Glaskowsky, N. A., D. R. Hudson, and R. M. Ivie, *Business Logistics*, 3rd ed. Wadsworth Publishers 1992 (ISBN 0534510353).
10. Handfield, R. B., and E. Z. Nichols, *Introduction to Supply Chain Management*. Englewood Cliffs, NJ: Prentice-Hall, 1998 (ISBN 0136216161).

11. Jones, J. V., *Companion to the Logistics Support Analysis Record.* LMA-TLM Press, 1998 (ISBN 0958725551).
12. Kasilingam, R. G., *Logistics and Transportation: Design and Planning.* Boston: Kluwer Academic Publishers, 1999 (ISBN 0412802902).
13. Lambert, D. M., J. R. Stock, L. M. Ellram, and J. Stockdale, *Fundamentals of Logistics Management.* New York: McGraw-Hill, 1997 (ISBN 0256141177).
14. Leenders, M. R. *Purchasing and Materials Management*, 10th ed. New York: McGraw-Hill, 1992 (ISBN 0256103348).
15. MIL-HDBK-59A, Military Handbook, *Computer-Aided Acquisition Logistic Support (CALS) Implementation Guide.* Washington, DC: Department of Defense (latest edition).
16. MIL-HDBK-502, Department of Defense Handbook, *Acquisition Logistics.* Washington, DC: Department of Defense (latest edition).
17. MIL-PRF-49506, Performance Specification, *Logistics Management Information.* Washington, DC: Department of Defense (latest edition).
18. MIL-STD-1840A, Military Standard, *Automated Interchange of Technical Information.* Washington, DC: Department of Defense (latest edition).
19. Nerseian, R. L., and G. B. Swartz, *Computer Simulation in Logistics.* New York: Quorum Books, 1996 (ISBN 0899309852).
20. Patton, J. D., *Logistics Technology and Management: The New Approach—A Comprehensive Handbook for Commerce, Industry, Government.* New York: Solomon Press, 1986 (ISBN 0934623023).
21. Pooler, V. H. and D. Pooler, *Purchasing and Supply Management: Creating the Vision.* London: Chapman & Hall, 1997 (ISBN 0412106019).
22. Robeson, J. F. (Preface) and W. C. Copacino (Editor), *The Logistics Handbook.* New York: Free Press, 1994 (ISBN 0029265959).
23. Stock, J. R., and D. M. Lambert, *Strategic Logistics Management*, 3d ed. Irwin Professional Publishers, 1992 (ISBN 0256088381).
24. Tilanus, B., *Information Systems in Logistics and Transformation*, 2d ed. New York: Elsevier Science, 1997 (ISBN 0080430546).
25. Tompkins, J. A., and D. A. Harmelink (eds.), *The Distribution Management Handbook.* New York: McGraw-Hill, 1993 (ISBN 0070650462).
26. Wood, D. F., D. L. Wardlow, P. R. Murphy, and J. C. Johnson, *Contemporary Logistics.* Englewood Cliffs, NJ: Prentice-Hall, 1999 (ISBN 0137985487).

SYSTEMS, SYSTEMS ENGINEERING, AND SYSTEMS ANALYSIS

Primary References

1. Blanchard, B. S., *System Engineering Management*, 2d ed. New York: Wiley, 1998 (ISBN 0471190861).
2. Blanchard, B. S., and W. J. Fabrycky, *Systems Engineering and Analysis*, 3d ed. Englewood Cliffs, NJ: Prentice-Hall, 1998 (ISBN 0131350471).
3. Defense Systems Management College (DSMC), *Systems Engineering Management Guide.* Fort Belvoir, VA: DSMC (latest edition).

Additional References

1. Andriole, S. J., *Managing Systems Requirements: Methods, Tools, and Cases.* Englewood Cliffs, NJ: Prentice-Hall, 1996 (ISBN 0070019746).
2. Belcher, R., and E. Aslaksen, *Systems Engineering.* Englewood Cliffs, NJ: Prentice-Hall, 1992 (ISBN 0138804028).
3. Checkland, P., *Systems Thinking, Systems Practice.* New York: Wiley, 1981 (ISBN 0471279110).
4. Martin, J. N., *Systems Engineering Guidebook: A Process for Developing Systems and Products.* Boca Raton, FL: CRC Press, 1996 (ISBN 0849378370).

5. Rechtin, E., and M. W. Maier, *The Art of Systems Architecting.* Boca Raton, FL: CRC Press, 1996 (ISBN 0849378362).

6. Sage, A. P., *Systems Engineering.* New York: Wiley, 1992 (ISBN 0471536393).

SOFTWARE AND COMPUTER-AIDED SYSTEMS

Primary Reference

1. Pfleeger, S. L., *Software Engineering: Theory and Practice.* Englewood Cliffs, NJ: Prentice-Hall, 1998 (ISBN 013624842X).

Additional References

1. Bass, L., P. Clements, R. Kazman, and K.Bass, *Software Architecture in Practice.* Reading, MA: Addison-Wesley, 1998 (ISBN 0201199300).

2. Boehm, B. W., *Software Engineering Economics.* Englewood Cliffs, NJ: Prentice-Hall, 1981 (ISBN 0138221227).

3. Humphrey, W. S., *A Discipline for Software Engineering.* Reading, MA: Addison-Wesley, 1995 (ISBN 0201546108).

4. Pressman, R. S., *Software Engineering: A Practitioner's Approach*, 4th ed. New York: McGraw-Hill, 1996 (ISBN 0070521824).

5. Sage, A. P., *Software Systems Engineering.* New York: Wiley, 1990 (ISBN 047161758X).

6. Sage, A. P., *Systems Management for Information Technology and Software Engineering.* New York: Wiley, 1995 (ISBN 0471015830).

7. Thayer, R., M. Dorfman, and S. C. Bailin (eds.), *Software Requirements Engineering.* New York: IEEE Computer Society, 1997 (ISBN 0818677384).

RELIABILITY ENGINEERING

Primary References

1. Ebeling, C. E., *An Introduction to Reliability and Maintainability Engineering.* New York: McGraw-Hill, 1996 (ISBN 0070188521).

2. Knezevic, J., *Reliability, Maintainability, and Supportability: A Probabilistic Approach.* New York: McGraw-Hill, 1993 (ISBN 0077076915).

3. Jones, J. V., *Engineering Design: Reliability, Maintainability and Testability.* New York: McGraw-Hill, 1988.

Additional References

1. Barlow, R. E., *Engineering Reliability (Statistics and Applied Probability.* Society of Industrial & Applied Mathematics, 1998 (ISBN 0898714052).

2. Ireson, W. G. (ed.), *Handbook of Reliability Engineering and Management.* New York: McGraw-Hill, 1996 (ISBN 0070127506).

3. Kales, P., *Reliability: For Technology, Engineering, and Management.* Englewood Cliffs, NJ: Prentice-Hall, 1997 (ISBN 0134858220).

4. Kececioglu, D., *Reliability Engineering Handbook,* Vols. 1 and 2. Englewood Cliffs, NJ: Prentice-Hall, 1991 (ISBN 013772294X).

5. Lewis, E. E., *Introduction to Reliability Engineering*, 2d ed. New York: Wiley, 1995 (ISBN 0471018333).

6. Lyu, M. R. (ed.), *Handbook of Software Reliability Engineering*. New York: McGraw-Hill, 1996 (ISBN 0070394008).

7. Musa, J. D., *Software Reliability Engineering: More Reliable Software, Faster Development and Testing*. New York: McGraw-Hill, 1998 (ISBN 0079132715).

8. O'Connor, P. D. T., D. Newton, and R. Bromley, *Practical Reliability Engineering*, 3d ed. Englewood Cliffs, NJ: Wiley, 1996 (ISBN 0471957674).

9. Palady, P., *Failure Modes and Effects Analysis*, private publcation, 1995 (ISBN 0945456174).

10. Pecht, M. (ed.), *Product Reliability, Maintainability, and Supportability Handbook*. Boca Raton, FL: CRC Press, 1995 (ISBN 0849394570).

MAINTAINABILITY ENGINEERING AND MAINTENANCE

Primary References

1. Blanchard, B. S., D. Verma, and E. L. Peterson, *Maintainability: A Key to Effective Serviceability and Maintenance Management*. New York: Wiley, 1995 (ISBN 0471591327).

2. Knezevic, J., *Systems Maintainability: Analysis, Engineering, and Management*. London: Chapman and Hall, 1997 (ISBN 0412802708).

3. Moubray, J., *Reliability-Centered Maintenance*, 2d ed. Industrial Press, 1997 (ISBN 0831130784).

Additional References

1. Nakajima, S., *Introduction to TPM: Total Productive Maintenance*. Productivity Press, 1994 (ISBN 0915299232).

2. Nakajima, S. (ed.), *TPM Development Program: Implementing Total Productive Maintenance*. City, State: Productivity Press, 1989 (ISBN 0915299372).

3. Niebel, B.W., *Engineering Maintenance Management*, 2d ed. New York: Marcel Dekker, 1994 (ISBN 0824792475).

4. Patton, J. D., *Maintainability and Maintenance Management*, 3d ed. Instrument Society of America, 1994 (ISBN 1556175108).

5. Smith, A. M., *Reliability-Centered Maintenance*. New York: McGraw-Hill, 1992 (ISBN 007059046X).

6. Willmott, P., *Total Productive Maintenance*. Boston: Butterworth-Heinemann, 1995 (ISBN 0750619252).

7. Wireman, T., *Developing Performance Indicators for Managing Maintenance*. Industrial Press, 1998 (ISBN 0831130806).

HUMAN FACTORS AND SAFETY ENGINEERING

Primary References

1. Bahr, N. J., *System Safety Engineering and Risk Assessment: A Practical Approach*. New York: Taylor & Francis, 1997 (ISBN 1560324163).

2. Chaffin, D. B., and G. B. J. Andersson, *Occupational Biomechanics*, 2d ed. New York: Wiley, 1991.

3. Diffrient, N., *Humanscale Series*. Boston: MIT Press, 1981.

4. Eastman Kodak, *Ergonomic Design for People at Work,* Vol. 1. New York: Van Nostrand Reinhold, 1983.

5. Salvendy, G. (ed.), *Handbook of Human Factors and Ergonomics*, 2d ed. New York: Wiley, 1997 (ISBN 0471116904).

Additional References

1. Kroemer, K. H. E., and H. J. Kroemer, *Engineering Physiology: Basis of Human Factors/Ergonomics*, 3d ed. New York: Wiley, 1997 (ISBN 0471287989).
2. Roland, H. E., and B. Moriarity, *System Safety Engineering and Management*, 2d ed. New York: Wiley, 1990 (ISBN 0471618160).
3. Sanders, M. S., and E. J. McCormick, *Human Factors in Engineering and Design*, 7th ed. New York: McGraw-Hill, 1992 (ISBN 007054901X).
4. Wickens, C. D., S. Gordon, and Y. Liu, *An Introduction to Human Factors Engineering*. Reading, MA: Addison-Wesley, 1997 (ISBN 0321012291).

ENGINEERING ECONOMY, LIFE-CYCLE COST ANALYSIS, AND COST ESTIMATION

Primary References

1. Fabrycky, W. J., and B. S. Blanchard, *Life-Cycle Cost and Economic Analysis*. Englewood Clffs, NJ: Prentice-Hall, 1991 (ISBN 0135383234).
2. Stewart, R.D., *Cost Estimating*, 2d ed. New York: Wiley, 1991 (ISBN 0471857076).

Additional References

1. Canada, J. R., W. G. Sullivan, and J. A. White, *Capital Investment Analysis for Engineering and Management*, 2d ed. Englewood Cliffs, NJ: Prentice-Hall, 1996 (ISBN 0133110362).
2. Fabrycky, W. J., G. J. Thuesen, and D. Verma, *Economic Decision Analysis*. Englewood Cllffs, NJ: Prentice-Hall, 1997 (ISBN 0133702499).
3. Grant, E. L., W. G. Ireson, and R. S. Leavenworth, *Principles of Engineering Economy*, 8th ed. New York: Wiley, 1990 (ISBN 047163526X).
4. Hicks, D. T., *Activity-Based Costing: Making It Work for Small and Mid-Sized Companies*, 2d ed. New York: Wiley, 1998 (ISBN 0471249599).
5. Ostwald, P. F., *Engineering Cost Estimating*, 3d ed. Englewood Cliffs, NJ: Prentice-Hall, 1992 (ISBN 0132766272).
6. Thuesen, G. J., and W. J. Fabrycky, *Engineering Economy*, 8th ed. Englewood Cliffs, NJ: Prentice-Hall, 1993 (ISBN 0132799286).

MANAGEMENT AND SUPPORTING AREAS

1. Boxwell, R. J., *Benchmarking for Competitive Advantage*. New York: McGraw-Hill, 1994 (ISBN 0070068992).
2. Brigham, E. F., D. A. Clark, and J. F. Houston, *Fundamentals of Financial Management*. Chicago: Dryden Press, 1997 (ISBN 003024434X).
3. Camp, R. C., *Business Process Benchmarking: Finding and Implementing Best Practices*. Quality Resources, 1995 (ISBN 0873892968).
4. Cleland, D. I., *Project Management: Strategic Design and Implementation*, 3d ed. New York: McGraw-Hill, 1998 (ISBN 007012020X).
5. Cleland, D. I., and R. Gareis, *Global Project Management Handbook*, 2d ed. New York: McGraw-Hill, 2006 (ISBN 0-07-146045-4).
6. Department of Defense (DOD), *Operating and Support Cost-Estimating Guide*. Washington: Office of the Secretary of Defense Cost Analysis Improvement Group (CAIG), OASD (PA&E), Room 2D278, The Pentagon, Washington, DC 20301, May 1992.

7. Defense Systems Management College (DSMC), *Program Manager*, Journal published bimonthly, DSMC, Attention: DSMC Press, 9820 Belvoir Rd., Ste 3, Fort Belvoir, VA 22060-5565 (ISSN 0199-7114).

8. Gordon, J. R., and S. R. Gordon, *Information Systems: A Management Approach*, 2d ed. Chicago: Dryden Press, 1998 (ISBN 0030224691).

9. Griffin, R.W., *Management*, 5th ed. Boston: Houghton Mifflin, 1995 (ISBN 0395731100).

10. Hesse, R., *Managerial Spreadsheet Modeling and Analysis*. Richard D. Irwin, 1996 (ISBN 0256215308).

11. Johnson, R. A., I. Miller, and J .E. Freund, *Probability and Statistics for Engineers*, 5th ed. Englewood Cliffs, NJ: Prentice-Hall, 1994 (ISBN 0137214081).

12. Kerzner, H., *Project Magement: A Systems Approach to Planning, Scheduling, and Controlling*, 6th ed. New York: 1997 (ISBN 0471288357).

13. Koontz, H., and H. Weihrich, *Essentials of Management*, 5th ed. New York: McGraw-Hill, 1990 (ISBN 007035605X).

14. Laudon, K. C., and J. P. Laudon, *Essentials of Management Information Systems*. Englewood Cliffs, NJ: Prentice-Hall, 1997.

15. Lewis, J. P., *Fundamentals of Project Management*. Amacom, 1995 (ISBN 0814478352).

16. Magrab, E. B., *Integrated Product and Process Design and Development: The Product Realization Process*. Boca Raton, FL: CRC Press, 1997 (ISBN 0849384834).

17. Ostrofsky, B., "Design, Planning, and Development Methodology," 14611 Carolcrest Drive, Houston, TX.

18. Thamhain, H. J., *Engineering Management: Managing Effectively in Technology- Based Organizations*. Englewood Cliffs, NJ: Wiley, 1992 (ISBN 0471828017).

APPENDIX D
PRODUCIBILITY DESIGN ASSESSMENT CHECKLIST

1. General Design Issues

a. Have alternative design concepts been considered and the simplest and most producible one selected?
b. Does the design exceed the manufacturing state of the art?
c. Is the design conducive to the application of economic processing?
d. Does a design already exist for the item?
e. Does the design specify the use of proprietary items or processes?
f. Is the item overdesigned or underdesigned?
g. Can a simpler manufacturing process be used?
h. Can redesign eliminate anything?
i. Is motion or power wasted?
j. Can the design be simplified?
k. Can parts with slight differences be made identical?
l. Can compromises and trade-offs be used to a greater degree?
m. Is there a less costly part that will perform the same function?
n. Can a part designed for another equipment be used?
o. Can weight be reduced?
p. Is there something similar to this design that costs less?
q. Can the design be made to secure additional functions?
r. Are quality assurance provisions too rigorous for design or functions?
s. Can multiple parts be combined into a single net shape?

2. Specifications and Standards

a. Can the design be standardized to a greater extent?
b. Can the design use standard cutting tools to a greater extent?
c. Is there a standard part that can replace a manufactured item?
d. Can any specifications be relaxed or eliminated?
e. Can standard hardware be used to a greater degree?
f. Can standard gages be used to a greater degree?
g. Are nonstandard threads used?
h. Can stock items be used to a greater degree?
i. Should packaging specifications be relaxed?
j. Are specifications and standards consistent with the planned product environment?

3. Drawings

    a. Are drawings properly and completely dimensioned?
    b. Are tolerances realistic, producible and not tighter than the function requires?
    c. Are tolerances consistent with multiple manufacturing process capabilities?
    d. Is required surface roughness realistic, producible ad not better than function requires?
    e. Are forming, bending, fillet and edge radii, fits, hole sizes, reliefs, countersinks, O-ring grooves, and cutter radii standard and consistent?
    f. Are all nuts, bolts, screws, threads, rivets, torque requirements, etc. appropriate and proper?
    g. Have requirements for wiring clearance, tool clearance, component space, and clearance for joining connectors been met?
    h. have all required specifications been properly invoked?
    i. Are adhesives, sealants, encapsulants, compounds, primers, composites, resins, coatings, plastics, rubber, moldings, and tubing adequate and acceptable?
    j. Has galvanic corrosion and corrosive fluid entrapment been prevented?
    k. Are welds minimal and accessible, and are the symbols correct?
    l. Have design aspects that could contribute to hydrogen embrittlement, stress corrosion, or similar conditions been avoided?
    m. Are lubricants/fluids proper?
    n. Are contamination controls of functional systems proper?
    o. Have limited life materials been identified, and can they be replaced without difficulty?
    p. Have radio frequency interference (RFI) and electromagnetic interference (EMI) shielding, electrical, and static bond paths been provided?
    q. Have spare connector contacts been provided?
    r. Are identification and marking schemes for maximum loads, pressure, thermal, nonflight items, color codes, power, and hazards on the drawings properly?
    s. Do drawings contain catchall specifications that manufacturing personnel would find difficult to interpret?
    t. Have all possible alternatives of design configuration been shown?

4. Materials

    a. Have materials been selected that exceed requirements?
    b. Will all materials be available to meet the required need dates?
    c. Have special material sizes and alternate materials been identified, sources verified, and coordination effected with necessary organizations?
    d. do design specifications unduly restrict or prohibit use of new or alternate materials?
    e. Does the design specify peculiar shapes requiring extensive machining or special production techniques?
    f. Are specified materials difficult or impossible to fabricate economically?
    g. Are specified materials available in the necessary quantities?
    h. Is the design flexible enough so that many processes and materials can be used without functionally degrading the end item?
    i. Can a less expensive material be used?
    j. Can the number of different materials be reduced?
    k. Can a lighter gauge material be used?
    l. Can another material be used that would be easier to machine?
    m. Can use of critical materials be avoided?
    n. Are alternate materials specified where possible?
    o. Are materials and alternates consistent with all planned manufacturing processes?

5. Fabrication Processes

a. Does the design involve unnecessary machining requirements?
b. Have proper design specifications been used with regard to metal stressing, flatness, corner radii, types of castings, flanges and other proper design standards?
c. Does the design present unnecessary difficulties in forging, casting, machining, and other fabrication processes?
d. Do the design specifications unduly restrict production personnel to one manufacturing process?
e. Can parts be economically subassembled?
f. Has provision been made for holding or gripping parts during fabrication?
g. Are expensive special tooling and equipment required for production?
h. Have the most economical production processes been specified?
i. Have special handling devices or procedures been initiated to protect critical or sensitive items during fabrication and handling?
j. Have special skills, facilities, and equipment been identified and coordinated with all affected organizations?
k. Can parts be removed or disassembled and reinstalled or reassembled easily and without special equipment or tools?
l. Is the design consistent with normal shop flow?
m. Has consideration been given to measurement difficulties in the production process?
n. Is the equipment and tooling list complete?
o. Are special facilities complete?
p. Can a simpler manufacturing process be used?
q. Have odd size holes and radii been used?
r. Can a fastener be used to eliminate tapping?
s. Can weld nuts be used instead of a tapped hole?
t. Can roll pins be used to eliminate reaming?
u. Do finish requirements prohibit use of economical speeds and feeds?
v. Are processes consistent with production quantity requirements?
w. Are alternate processes possible within design constraints?

6. Joining Methods

a. Are all parts easily accessible during joining processes?
b. Are assembly and other joining functions difficult or impossible due to lack of space or other reasons?
c. Can two or more parts be combined into one?
d. Is there a newly developed or different fastener to speed assembly?
e. Can the number of assembly hardware sizes be minimized?
f. Can the design be changed to improve the assembly or disassembly of parts?
g. Can the design be improved to minimize installation or maintenance problems?
h. Have considerations for heat-effected zones been considered when specifying a thermal joining process?

7. Coating Materials and Methods

a. Are protective finishes properly specified?
b. Has corrosion protection been adequately considered from the standpoint of materials, protective measures, and fabrication and assembly methods?

c. Have special protective finish requirements been identified and solutions defined?
d. Can any special coating or treating be eliminated?
e. Can pre-coated materials be used?

8. Heat Treating and Cleaning Processes

a. Is the specified material readily machined?
b. Are machining operations specified after heat treatment?
c. Have all aspects of production involving heat treatment and cleaning processes and their interaction with other production areas been reviewed?
d. Are heat treatments properly specified?
e. Are process routings consistent with manufacturing requirements?

9. Safety

a. Have static ground requirements been implemented in the design?
b. Have necessary safety precautions been initiated for pyrotechnic items?
c. Have RFI and EMI requirements been implemented in the design?
d. Have necessary safety requirements for processing materials, such as magnesium and beryllium copper, been considered?

10. Environmental Requirements

a. Have adequate provisions been included to meet OSHA or other special environmental or health hazard issues?
b. Have adequate provisions been made to contain or control use of hazardous materials or chemicals in the manufacturing process?
c. Have adequate provisions been mate to contain or control hazardous byproducts of the manufacturing process?
d. Have alternate designs been considered to reduce or eliminate environmental issues?
e. Have all affected organizations been informed of possible environmental issues?

11. Inspection and Test

a. Are inspection and test requirements excessive?
b. Is special inspection equipment specified in excess of actual requirements?
c. Is the item inspectable by the most practical method possible?
d. Have conditions or aspects anticipated to contribute to high rejection rates been identified and remedial action initiated?
e. Have required mock-ups and models been provided?
f. Are special and standard test and inspection equipment on hand, calibrated, proofed, and compatible with drawing requirements?
g. Are master and special gages complete?
h. Have nondestructive testing techniques been implemented?
i. Have adequate provisions been made for the checkout, inspection, testing, or proofing if functional items in accordance with operational procedures?
j. Is nonstandard test equipment necessary?
k. Are all aspects of acceptance contained in the inspection and test planning?

Producibility Issues for Design Reviews

1. Materials
   a. Are materials, including alternatives, off the shelf purchases?
   b. Have materials been standardized o the maximum extent possible?
   c. Have estimates lead times for the delivery of materials been established?
   d. Are materials lead times acceptable?
   e. Do material properties exceed the requirements?
   f. Have all special material needs been identified and verified?
   g. Can a lower cost material be used?
   h. Are materials and alternatives consistent with the most efficient manufacturing process?
   i. Have the proper design specifications been used to specify material properties after the manufacturing process?
   j. Have material producibility trade-offs caused deterioration of the minimum design requirements?

2. Manufacturing Processes
   a. Will planned manufacturing technology developments be available?
   b. Have all production feasibility risk analyses been completed?
   c. Are plans for proof testing critical processes adequate?
   d. Are plans for proof testing tooling adequate?
   e. Are plans for proof testing test equipment adequate?
   f. Do planned processes have necessary tolerance capabilities?
   g. Does the design create unnecessary difficulties in forging, casting, machining, and other processes?
   h. Are materials and quantities consistent with planned processes?
   i. Are production processes and personnel available?
   j. Has necessary tooling (jigs and fixtures) been adequately considered?
   k. Have the most economical processes been specified?

3. Design Process
   a. Have component and materials standardization been maximized?
   b. Have the effects of trade-off studies for producibility been reflected in the design?
   c. Have critical materials (types and quantities) been minimized?
   d. Have constraints on fabrication and assembly been minimized?
   e. Has the use of existing or new industrial resources been proven?
   f. Have adequate management initiatives and organization been established?

APPENDIX E
SAFETY DESIGN REQUIREMENTS

1 GENERAL REQUIREMENTS

1.1 Criteria

In addition to applicable federal regulatory standards, all machines shall be designed in accordance with the provisions of the criteria contained herein, and the principles and concepts in the standards identified below. Due to periodic changes in the respective standards, use most current one.

- National Fire Protection Association: Electrical Standard for Industrial Machinery, NFPA 79.
- National Fire Protection Association: National Electrical Code, NFPA 70.
- National Fire Protection Association: National Fire Alarm Code, NFPA 72.
- National Fire Protection Association: Life Safety Code, NFPA 101.
- National Fire Protection Association: Installation of Sprinkler Systems, NFPA 13
- Underwriters Laboratories, Inc.: Standard for Information Technology Equipment, UL-60950. (Note: This standard is to replace Safety of Electrical Business Equipment, UL-1950. See information issued with standards concerning application dates.)
- Underwriters Laboratories, Inc.: Safety-Related Software, UL-1998.

2 NOISE

When measured with an American National Standards Institute Specifications For Sound Level Meters, ANSI S1.4, Type 1, sound level meter set to A-weighting and slow response, the emitted noise level shall not exceed 80 decibels (dB) for the following conditions:

- Any equipment operating condition or mode excluding the required audible start-up alarm signals.
- Measurements shall be taken one (1) foot (30 cm) from the external platform (geometric plane) surface, at a height of five (5) feet (152 cm) from the floor or other walking surface, and circumscribing the machine. Due to different postural or working positions, e.g. sitting in a chair, the height from the floor may be changed to reflect a different expected operator's ear location.
- Pure tones shall be at least 10 decibels less in its octave band than the measured A-weighted noise level for the equipment at the same location while operating under any of its normal conditions.

3   EMERGENCY STOP SWITCHES

3.1   Functions of Emergency Stop Switches

The functions of an emergency stop switch shall be as follows:

- Setting of an emergency stop switch shall immediately remove primary power to all mechanical parts that are in motion or could be set into motion, and stop all such respective parts.
- Respective mechanical parts shall remain de-energized and stopped as long as an emergency switch is set.
- Setting of an emergency stop switch shall result in the removal of primary power by one or more, as appropriate by design, hardware devices for disconnecting power.

3.2   Accessibility of Emergency Stops

An emergency stop switch or switches shall be accessible at the following locations:

- At the operator's control console(s) or panel(s),
- At all other location around the platform edges unless specifically exempted, and
- Where an individual can directly or indirectly reach into or within an area of the machine where there are or might be unguarded safety and health hazards. This includes situations where:
  - a) Operators or supervisors might reach into or around covers of areas with moving or rotating machine parts,
  - b) Operators or supervisors are expected to access normally guarded areas to clear jams or perform other designated functions, and
  - c) Maintenance personnel must work within such an area with an interlock switch defeated.

3.3   Location of Emergency Stop Switches

Emergency stop switches shall be located so that they are:

- Within four feet (122 cm) of any point around the periphery of the equipment
- Not more than two feet (61 cm) from the platform edge, and
- Have free and clear access in both vertical and horizontal directions at all times

3.4   Emergency Stop Resets

The reset for manually-activated emergency stop switches shall be at the switch itself or, if not, shall be immediately adjacent to the same switch and at no other location.

3.5   Emergency Stop Circuits

Emergency stop circuits shall be hardwired.

3.6   Restart after Stops

The resetting of an activated emergency switch, clearance of a jammed transport belt, or remedying conditions that originally caused a machine to stop shall not result in the automatic restarting of the machine.

4 MACHINE GUARDS

4.1 Machine Guarding

One or more methods of machine guarding shall be provided to protect operators and maintenance personnel in the machine area from operational hazards such as, but not limited to, those created by points of operation, in-running nip points, rotating parts, sharp corners and edges, protruding parts, moving belts, exposed electrical terminals greater than 30 volts Root Mean Square or RMS (42.4 volts peak or dc), high amperage-low voltage sources, excessive leakage currents, unsecured overhead objects, etc., when such hazards are located within 7 feet (213 cm) of the floor or work platform.

Notwithstanding the use of interlocks on access panels, additional guards shall be required when the accessed area includes more that one type of potential safety hazard. E.g., high-voltage electrical terminals that require maintenance monitoring and that are located adjacent to moving motor driven belts, gears, or pulleys.

This includes, for example, any of the following situations:

- Where operators or supervisors might reach into or around covers into areas with moving or rotating machine parts,
- Where operators or supervisors are expected to access normally guarded areas to clear jams or perform other designated functions, and
- Where operators or maintenance personnel can be working within such an area with an interlock switch defeated.

4.2 Corners, Edges, and Protrusions

All nonfunctional corners, edges, and protrusions that can cause cuts, punctures, contusions, or other safety concerns, shall be rounded, guarded, or eliminated by other means.

4.3 Hinges for Covers and Door Panels

All covers and door panels providing any operator or any maintenance access, including normal operation, shall be attached to the equipment with hinges.

Note 1

- Pin hinges which allow removal of door or cover for maintenance are acceptable.

Exceptions to Cover and Door Panel Hinges: Covers and door panels do not require hinges if all of the following conditions are met and can be demonstrated.

- Interior area is fully accessible from the adjacent walking surface.
- The interior space, components, and any required maintenance access is available through an immediately adjacent alternate cover or from another direction of approach.
- There is a safe place in the immediate area for temporary storage of the covers or door panels when removed for maintenance.
- Is self supporting during reinstallation and refastening.
- Width is not more than 30 inches.
- Height not more than 36 inches.
- Weight not more than 20 pounds.
- Removal is not required more than twice per year.

4.4 Shaft End Caps

End cover caps shall be installed for any exposed shaft end that includes protrusions, such as, but not limited to set screws, above the shaft surface.

5 EQUIPMENT CRITERIA

5.1 Uninterruptible Power Supplies, Identification

Uninterruptible power supplies used in support of computer type hardware shall be identified in some manner such as 1/4 inch high letters or other form of markings that can be easily read to indicate the nature of the component.

5.2 Uninterruptible Power Supplies, Safe Shut Off Procedures

All equipment equipped with an uninterruptible power supply used to provide continuous electrical power to computer related hardware in case of such events as the lost of facility electrical power, local power, or unintentional removal of equipment power shall be labeled with notification of the use of the uninterruptible power supply and the manner in which it can be intentionally and safely shut down for equipment lockout purposes. Such labels shall be mounted in the vicinity of the primary electrical disconnect and be legible to an individual standing approximately 20 feet away from it. Such procedures shall be included in the written maintenance procedures.

5.3 Material Safety Data Sheets

Material Safety Data Sheets (OSHA Form 174 or equivalent pursuant to 29 CFR 1910.1200) shall be provided for any specific chemical product, including inks, required for routine use or maintenance of the equipment. Additional safety and health data may be required if there are other concerns that are not addressed satisfactorily by the MSDS information.

5.4 Use of Flammable or Combustible Liquids

When the routine use of a combustible or flammable product is required, the Buyer shall be provided with a safety and health review of that product. It shall address such concerns as, but is not limited to, the following items: safe storage, transportation handling requirements, toxicity, flammability characteristics, and compliance with Art 500 of the National Electrical Code.

5.5 Training Courses and Operations Manuals

All training courses and operations manuals shall include information on potential safety and health hazards, if any. They shall also include the steps, or manner of use, to be observed by operating and maintenance personnel to ensure safety. This information shall be identified as to whether it is a hazard to operating personnel, maintenance personnel, or both.

5.6 Safety labels

To the extent that the use of existing safety labels are applicable to the equipment, such labels shall be a part of the equipment.

5.7 Equipment Identification

All major equipment shall be identified by 1 inch or larger characters that include the manufacturer's name when appropriate, the manufacturer's or buyer's designated functional or identification name of the equipment, and clearly designated model number in a prominently visible external surface location.

If a manufacturer's nameplate appears on the equipment, it should not be physically covered, removed or painted over unless prior approval by the buyer safety contact is acquired.

5.8 Power Requirements Label

Where applicable and appropriate, equipment shall be labeled with the electrical power (phases, voltage, and current) requirements.

5.9 Stools—Glides

When stools are provided as workstation furniture, only glides shall be provided with stools due to their higher sitting position above the floor. Casters are not acceptable.

5.10 Equipment Controls

The use and operation of equipment is generally identified as being by an "operator", supervisor, or maintenance personnel. Except for primary duty work requiring manual data entry, address reading or verification, general processing equipment designed to be operators controlled shall not require the use or interaction with computers, associated keyboards, or monitors for its operation including identification of jam locations. Supervisors have access to computer keyboards and monitors for necessary supervisory functions. Where appropriate and designated for maintenance only, access to computer keyboards and other information entry devices shall be physically limited to maintenance personnel only.

APPENDIX F
HUMAN FACTORS ENGINEERING DESIGN REQUIREMENTS

1 SEATED WORKSTATIONS

1.1 Normal Seating Device

A seated workstation should be provided for situations where continuous data entry operations occur and the operator is not required to move away from the workstation. Examples of such workstations are the office automation, remote bar coding, and computerized forwarding operations. The workstation is designed to provide an upright posture with sufficient back support, foot support and wrist support (if requested by the operator). The design should ensure that discomfort due to repetitive motion is minimized. Similarly, visual fatigue and discomfort is minimized through the design of the VDT and the work surface surroundings.

The system shall be designed to minimize user errors and fatigue. According to the accepted ergonomics practices, an integrated approach is taken toward the design where the requirements for VDT, keyboard, software, keying console are provided.

1.1.1 Visual Display Terminal (VDT). Below are the minimum requirements for the design of the VDT for Luminance:

- Screen luminance: 75 cd/m^2–200 cd/m^2
- Character luminance: 10 cd/m^2–20 cd/m^2

The display image shall be a non-interlaced display using black characters on an 'paper' white background with no black borders on the screen. This shall be a part of the hardware design of the VDT.

1.1.1.1 Adjacent Surfaces Surfaces adjacent to the scope shall have a dull matte finish.

1.1.1.2 Reflection and Glare Reflection and glare should be prevented through the use of a hood, optical coatings or filter control.

1.1.1.3 Finish and Luminance of Surrounding Area Panel surfaces adjacent to the VDT must have a dull matte finish, and the luminance range must be between 10 to 100 % of the screen background.

1.1.1.4 Phosphor The display phosphors must have short to medium persistence, be newspaper white (fluorescent and phosphorescent), and have a positive display image (black characters on a newspaper white background). Additionally, the phosphor must have color coordinate values between 0.265 and 0.355 for the X value, and 0.295 and 0.395 for the Y value.

1.1.1.5 Persistence Persistence should be from short to medium

Short	1μ sec. to 10μ sec.
Medium to Short	10μ sec. to 1 msec.
Medium	1 msec. to 100 msec.

1.1.1.6 Refresh Rate   The refresh rate must be at least 68 HZ for non-interlaced (at least 100 HZ for interlaced) with no perceivable flicker on positive display in focal or peripheral vision.

1.1.1.7 Jitter   Erratic movement or sweep traces on VDT screens must be diminished so that they are not detectable by the user. As an example for the operators using 80 cd/m2 bright screens the physical jitter at 10 HZ should be less than 15 seconds of arc.

1.1.1.8 Hand Capacitance Effects   Aluminized backing of the screen must be used for situation where the user's hand comes close to the screen.

1.1.1.9 Burning of Screen   Anti-burn techniques must be used to eliminate burning of the screen.

1.1.1.10 Geometric Distortion   Variations in the geometric location of a picture element shall be equal to or less than 0.0002 mm per mm of viewing distance.

1.1.1.11 Screen Shape   Screen shape must be rectangular.

1.1.1.12 Useful Screen diameter   The useful screen diameter must be a minimum of 15 inches on the diagonal.

1.1.1.13 Display Composition   Controls must be provided for adjusting illumination and contrast.

1.1.1.14 Adjustability   The VDT must be tiltable/adjustable along its vertical and horizontal axes.

1.1.1.15 X-Ray Emission   DHEW Rules 21 CFR Subchapter J—Part 1000–1030 must be followed.

NOTE: Center for Devices and Radiological Health (CDRH) requires that VDTs with more than 0.4 mR/H should be tested by the Center for X-ray emission. CDRH requires all imported terminals to be tested by the Center for X-ray emission.

1.1.1.16 Image Display Magnification Factor   The magnification factor must be between 1.30 and 1.50.

1.1.1.17 Image Resolution on Display Hardware   Image resolution must be 200 dpi.

1.1.1.18 Image Grayscale   Minimum of 4 levels of Gray (2 bits)

1.1.1.19 Electronic Signal to Noise Ratio   This criterion refers to the dots (noise) on the images presented to the operators. The ratio shall not exceed 35 dB.

1.1.2 Alpha-Numeric Text Data Entry Screens.   Listed below are the requirements for alphanumeric data entry screens.

1.1.2.1 Legibility Requirements   Characters must be designed so as to avoid 'look alike' pairs which might be confused with one another (e.g. B, 8 and 5; O and 0 (zero); Z and 2)

1.1.2.2 Dot Matrix   The text dot matrix must be no less than 7×9.

1.1.2.3 Character Height   Upper case characters must be no less than 3/16" in height.

1.2   Keyboard

The ergonomic keyboard requirement must provide for rapid and accurate unobstructed visual search and data entry. The keyboard design must minimize operator error rates, memorization and reach requirements.

Layout and Configuration: The keyboard must have a QWERTY keypad. It must have numeric keys above the alpha keypad and a minimum of 15 single stroke programmable function keys. The distance between the function keys and the top row of alpha keypad must not exceed one inch.

1.2.1 Slope.   Slope of the keyboard must not exceed 12 degrees.

1.2.2 Detachable Keyboard.   Keyboard must be detached and have adjustable slope in order to provide the operators with freedom to adjust the keyboard to a desirable position.

1.2.2.1 Operator feedback   The keyboard must provide the operator feedback to include information that the key was pressed, and where applicable, the next operation may be initiated.

1.2.2.2 Color and Reflection   The keyboard panel and the key surfaces must be dull matte finished. The use of dark color is prohibited and the reflectance value must be between 10% and 16%.

1.2.2.3 Resistance and Displacement   For these requirements see Table 1.

1.2.2.4 Dimension and Separation   For these requirements see Table 2.

1.2.2.5 Keytop Label Dimensions   For requirements on keytop label dimensions see Table 3.

1.2.2.6 Key Titles   Key title must represent precisely the function it invokes.

TABLE 1   Keyboard Key Resistance

	Numeric	Alpha-Numeric
Minimum	0.9 oz.	0.9 oz.
Maximum	5.3 oz.	5.3 oz.

TABLE 2   Keyboard Key Displacement

	Numeric	Alpha-Numeric
Minimum	0.03 in.	0.05 in.
Maximum	0.19 in.	0.25 in.

TABLE 3   Keyboard Key Dimensions and Separation

	Function	Numeric	Alpha-Numeric
Key top	$1/2$ in. $\times$ $1/2$ in.	$1/2$ in. $\times$ $1/2$ in.	$7/8$ in. L $\times$ $1/2$ in. W
Center-to-center	$3/4$ in. $\times$ $3/4$ in.	$3/4$ in. $\times$ $3/4$ in.	1 in.
Separation	$1/4$ in. $\times$ $1/4$ in.		$1/2$ in.

TABLE 4   Keyboard Key Dimensions and Separation

Font height	$1/8$ in.
Font width	$1/16$ in.
Space between letters	$1/32$ in.
Space between lines	$1/16$ in.

1.3   Workstation Dimensions

The workstation must provide maximum adjustability and comfort for the operator. The workstation shall accommodate the 5th percentile female and the 95-percentile male of the US adult population. At a minimum the following requirements must be met.

1.3.1 Keying Console.   To provide the operators with maximum adjustability, the following dimensions shall be observed in the design of the video encoding workstation:

- Viewing distance must not exceed 21".
- Keyboard height (middle row to floor), adjustable from 27.6"–39.5".
- Screen center above floor adjustable from 35.4"–45.3".
- Screen backward inclination to horizontal plane, adjustable from 88°–105°

- Minimum leg clearance must be 27" (height from the bottom edge of the console).
- Minimum legroom (edge of the console to vertical panel of the workstation) must be 20".
- Minimum knee well entry width under the console must be 24".
- Minimum width for the workstation must be 48".
- Minimum height must be adjustable between 25 to 30" in order to accommodate the 5 to 95 percentiles of the operators.
- Minimum workstation depth must be 38".
- Surface of the workstation must be dull matte finish and the use of dark colors must be avoided.
- Display terminal must be positioned in such a way that the EYE LEVEL SITTING HEIGHT will be adjusted between 42 and 52".

1.3.2 Chairs. The chairs for the video-encoding operators must be armless.

Seat specifications:

Height	Adjustable from 15 to 21 in.
Width	18 in.
Angle	Adjustable from 4° to 6°

Upholstery must be porous, conduct body heat, no vinyl, not slick and must be replaceable padding must be latex. The thickness of the padding must be one inch.

Backrest:

Horizontal adjustability
Vertical adjustment in relation to seat from 115° to 120°
Height: 20 inches above seat surface
Width: 14 1/2 inches
Thickness of padding: at least 3/4 inch
Upholstery the same as the seat.

The profile of the backrest must have a "lumbar cushion" to support the lumbar area of the back.

Legs and Base:

A chair must have five legs with casters.
The diameter of the base must be 18 inches.

1.3.3 Foot Rests. Adjustable footrests must be provided and have an adjustability range of 3 to 6 1/2 inches.

1.3.4 Wrist Rests. Wrists rests shall be provided and meet the following specifications:

Height	1 7/8 in.
Width	20 in.
Length	11 7/8 in.
Keyboard angle (flat)	10°
Keyboard angle (legs engaged)	13°
Home row key height	1 5/8 in.
Pad length	18 3/8 in.
Pad depth	2 7/8 in.
Pad height (max)	1 1/4 in.
Pad height (min)	5/8 in.

The keyboard angle with the keyboard legs engaged and not engaged should fall into the preferred range of 10 to 20 degrees.

1.3.5 Anti Static Device. An anti-static mat should be provided at each workstation to prevent user generated static from interfering with the system. The mat should have sufficient friction on the surface coming into contact with the carpeting to prevent the mat from curling up or moving around.

1.4 Ergonomic Standards for Screens and Messages

Ergonomic software standards must provide for the consistency of the proposed design, reduced memorization, improved accuracy, and minimized transaction time for the user.

1.4.1 Screen Layout. This section describes the rules for designing the static layout of video display terminal (VDT) screens.

1.4.1.1 General Rules All screens must be 78 characters wide, and the information will be displayed beginning in column 2 and ending in column 79. The first three lines of each screen must be used for title and screen identification data. The second line of the screen must be blank and the third line must be dashes.

1.4.1.2 Titles The title must appear in upper case letters on the top line at the far left corner of the screen. Screens with identical titles, but different input data fields, must be further identified by a subtitle. This must appear right after the title and must be preceded by a dash.

1.4.1.3 System Date The date will be displayed at the upper right corner of the screen, on the same line as the title. The system date will be displayed as MM/DD/YY, and the system date will begin in column 64.

1.4.1.4 Identification Numbers Each screen must be assigned a six-character screen identifier. This six-character code must be placed at the upper right corner of the screen, on the same line as the title. The screen ID must begin in column 73. The first three positions indicate system, while the last three positions indicate the number of screens within the sequence.

1.4.1.5 Grouping Captions and variable data that serve primarily to identify the purpose and/or subject of the remaining data being entered or displayed on a screen must be grouped together. All information in an identification group must be presented in upper case letters. For screens with different identification data, more than one identification group must be developed. As an example, the zone information should be included in the identification block.

Explanatory phrases in this area must have the first letter of each word capitalized.

1.4.1.6 Error Messages And Prompts A message area must be reserved at the bottom of each screen. This area will occupy the lower left corner of the screen in the first 52 characters of the bottom four lines. For documentation purposes, a message is enclosed in brackets when displayed on the screen. The left bracket is located in column 1 and the right bracket is located to the right of the message and may not be located farther right than column 52.

ERROR MESSAGES: Messages developed by the system to designate a variety of error types. Error messages must be displayed immediately after an error occurs. The system should not allow the operator to proceed unless the error is corrected.

PROMPTS: Messages from the system instructing the user how to proceed. All messages, including prompts and error messages, must be in upper case letters.

1.4.1.7 Field Captions All words in field captions must have an initial capital letter.

1.4.1.8 Abbreviations Periods must not be used after abbreviations.

1.4.1.9 Display Fields Display fields are enclosed in brackets for documentation purposes. A bracket must appear in place of the blanks that immediately precede and follow the field in question. For fields that are set to a non-blank (or non-zero) the initial value are enclosed by brackets for documentation purposes.

1.4.1.10 Menu Layout Rules (If applicable)

- Menu screens shall be used when more than three choices are available to the operator.
- The input field for entering a menu choice must be located above the list of choices.
- Menu options must be ordered from top to bottom and left to right according to their frequency of occurrence. The most frequently used item will be the first item on the menu.
- A menu with 10 or fewer selections should, to the extent possible, have all of its selections arranged in one column.
- Menus with more than 10 selections should use multiple columns.
- When the selection fields are arranged in more than one column, a minimum of five blank spaces must be allowed between adjacent columns.
- For every selection, the selection number must be followed by a period, one space, and the selection description.

1.4.1.11 Selection Data Entry Fields

- Selection Data Entry Fields will be distinguished from other data entry fields by being enclosed in the symbols "< >."
- The selection caption heading must be followed immediately by a colon.
- There must be one blank space between the colon and the symbol "<".
- In instances where the selection items follow the selection data entry field, there must be one blank space between the symbol " >" and the selection items.

1.4.1.12 Normal Data Entry Screen Layout Rules

- Whenever possible, screen fields must be displayed in the same order as they appear on printed forms.
- Entry fields must be located to the right of the field caption after the colon.
- When a number of related fields appear on the same screen, they must be aligned vertically.

1.4.1.13 Special Layout Rules for Tabular Data

- Tabular data must have a heading for each column.
- To the extent possible, this heading will occupy a single line.
- Column headings must appear with an initial capital letter.
- The first line below the heading must consist of dashes defining the maximum limits of data or the full width of the caption, whichever is greater.
- The column heading must be centered in relation to the dashes.
- Tabular data must begin on the line immediately below the dashes.
- Data must be displayed so that the operator's eyes sweep from left to right in reading column entries.
- When more than seven lines of non-numeric tabular data are displayed on the screen, every sixth line must be blank. When there are two or more columns to be viewed by the operator every second line must be blank.
- Five blank spaces must be allowed between the adjacent columns.

1.4.1.14 Screen Overlay Rules (if applicable) A screen that is designed to overlay another screen without erasing the entire contents of the previous screen must display the entire previous screen. The additional material will follow the information displayed on the screen.

Only data fields on the latest screen must be enabled for input. If a field can be entered or modified on two screens, one of which overlays the other, the field must be shown on both screens in the same location.

1.5 Display Functions

This section describes the dynamic behavior allowed for screens.

- Fields that are not always needed in the interaction of the screen must be implemented as "pop-up" fields. They are to be displayed only when they are needed.
- If a "pop-up" field is used, it should not obscure other portions of the interface that the user may need to refer to to complete a task.
- Default values and informational messages must be displayed before the first operator input.
- Where it is necessary to guide the operator through the system, prompting messages must be used.
- As much as possible, the information must be displayed to the operator in a selection format.
- Prompting messages must be kept to a minimum.
- As much as possible, related data must be displayed in one screen to avoid extra keystrokes for the operator.
- Error messages must be brief, direct, and non-threatening.
- Error messages must instruct an operator how to correct an error.

- Error messages must be distinguished from other messages by being displayed in reverse video. (Only error messages are to be displayed in this way, when no other display options are available.)
- All messages must be in upper case letters.
- Messages must be removed or overwritten once they are no longer needed.

1.6   Operator Control Interface

This section describes the use of the Control function, and the Applications function keys, which the operator can use to direct the system. The keys that make up each of these groups and their functions are described below. The assignment of the function keys and their labeling must be reviewed and approved by the Buyer ergonomics staff.

1.6.1 Control Function Keys.   These keys are used to edit or act upon the contents of the current screen. All screens must support the following keys and their standard functions:

EXIT (color coded red):
This function key is used to exit a completed zone or to return to the first screen of a program.

CLEAR ENTRY (color coded red):
This function key will erase the value in the current field and allow the operator to re-enter a correct value.

PREVIOUS SCREEN (if applicable):
This function key is used when multiple screens are required to display the necessary information it allows the operator to view the previously displayed screen of information.

PREVIOUS ENTRY (if applicable):
This function key moves the cursor to the previously entered field.

ENTER:
This function key indicates that the operator has finished data entry to a screen field and notifies the system to process the information in the field.

REPEAT:
This function key is used when the operator must enter the same information as that of the immediately preceding address.

HELP:
The function key provides instructional information to the operator on the current program.

CORRECT (color coded yellow):
The function key erases the address entered on the screen and allows the operator to enter the correct address.

1.6.2 Applications Function Keys.   These keys must be used to select the next input function. They are tailored to the specific data entry task (usually a program) being performed. In general they have the following characteristics:

- They cause the system to advance to a new screen.
- They have a consistent function within a program.
- They are enabled or disabled as needed on a screen-by-screen basis.
- They are supported by specialized logic within each application.

These following keys are highly recommended, since they reduce the keystrokes by the operator.
Eight SCF dedicated function keys are RECOMMENDED where the eight most frequently used SCF are assigned to these keys. Numeric values of each SCF must be printed on the key.

Additional application function keys are:

- LETTER
- FLAT
- PARCEL

1.6.3 Use of Language.   This section describes the use of language for creating screen titles, field captions, and messages.

1.6.3.1 Titles and Subtitles   Titles and subtitles must be brief and identical to the ones used for ADRS training materials and videotapes.

1.6.3.2 Terminology   Industry standard terminology will be used unless new terminology is required for the development of new technology applications.

1.6.3.3 Abbreviations   When abbreviations are used, they must follow the Buyer standard abbreviations.

1.6.3.4 New Abbreviations   The Buyer Contracting Officer must approve new abbreviations.

1.6.4 Screen Messages.   The following requirements provide guidance for the development of error messages. The major objective is to ensure that all ADRS messages are consistent in style and structure, that they use the same words to communicate common meanings, and that they provide the specific information required to correct an operator's entry.

1.6.4.1 Specificity   Messages must be specific and reflect the reason they were displayed.

1.6.4.2 Practicality   Messages must be designed to guide users through the system, to inform them of errors and/or problems within the system, and to specify how to correct the errors.

1.6.4.3 Active Voice   Messages must be direct and in active voice. That is, the subject of the sentence must perform the action denoted by the verb.

1.6.4.4 Sentence Format   Error messages must be complete sentences followed by a period.

1.6.4.5 Character Limit   Error messages should not exceed 50 characters each.

1.6.4.6 Longer Messages   When longer messages are necessary, either make two sentences of less than 50 character each, or divide the sentence into two lines with less than 50 character each.

1.6.4.7 Omission of Articles   Articles must be omitted in error messages.

1.6.4.8 Consistent Diction   Use words consistently in messages. Acceptable terms that must be used in preparing error messages and the synonyms they replace include:

Acceptable	Not Acceptable
Select	Choose
Inform	Call
Modify	Change
Press	Hit, Push
Correct	Valid, legal, right
Find	Search, locate
Retain	Hold
Require	Need
View	See
Delete	Erase
Error	Mistake

1.6.4.9 Single Words   Single words must be used in place of phrases. Examples of some acceptable single terms to be used and the synonyms they replace include:

Acceptable	Not Acceptable
Enter	Key in
Return	Get Back, Go Back
Complete (a form)	Fill in (a form)
Exceed	Greater than

1.6.4.10 Prefixes Prefixes must not be used. Examples of acceptable terms to be used and synonyms they replace include:

Acceptable	Not Acceptable
Not Correct	Incorrect
Enter Again	Reenter
Not Authorized	Unauthorized

1.6.4.11 Contractions Verbs must not be abbreviated. Verb forms that must be used include:

Acceptable	Not Acceptable
Will Not	Won't
Do Not	Don't
Is Not	Isn't
Was Not	Wasn't
It Is	It's

1.6.4.12 Negation Avoid using NO for negation since this word also denotes an abbreviation for "number."

1.6.4.13 Abbreviations Avoid using abbreviations.

1.6.4.14 Sequential Selection Code Error Messages Error messages for sequential selection codes must follow one of the formats given below:

Sequential selection codes of three choices or less:

- SELECTION CODE MUST BE 1 OR 2.
- SELECTION CODE MUST BE 1, 2, OR 3.

Sequential selection codes exceeding three choices:

- SELECTION CODE MUST BE A THRU L.
- SELECTION CODE MUST BE 1 THRU 4.

Selection codes that are not sequential:

- NUMBER OF MONTHS MUST BE 3, 6, 9, OR 12.

1.6.4.15 Standard Field Formats Field formats must be shown exactly as given below.

Key: A = Alpha Characters
      X = Alpha/Numeric Characters
      9 = Numeric Characters

1.7 General Software Requirements

General software ergonomic requirements for address formats, use of numbers, use of prompts, and recovery and response time are listed below.

1.7.1 Numerals. When listing a series of choices, use numbers, starting with one, not zero. Always place a period after an item selection number.

1.7.2 Design of Prompts. The design of the prompts must follow the same requirements provided for messages.

1.7.3 Recovery Capabilities. The system must have recovery capabilities that allows a user to undo the previous command, as well as immediately return to the specific location where the operator was interacting with the system prior to system difficulty.

1.7.4 Software Design Criteria. Screens should be designed to accomplish the following:

- Minimize operator effort and operator memorization.
- Minimize operator frustration.

- Maximize use of operator habit patterns.
- Maximize tolerance for human differences.
- Maximize tolerance for environmental changes.

1.7.4.1 Displayed Information and System Response   The information displayed to an operator, such as symbols, abbreviations, displays codes, and alarms, shall be limited to that necessary to perform specific actions or to make decisions.

1.7.4.2 Program Failures   Programs should be designed so that a failure resulting from the computer program shall be distinguishable from equipment failure. Further, the program should allow for the orderly shutdown of operations and the retention of all transactions processed prior to system shutdown. The system will provide prompt, lucid, non-threatening error messages that sufficiently explain the nature of the problem in terms understandable to the user. The system shall also provide, upon request, additional and more elaborate explanation.

1.7.4.3 Symbols   Where symbols are used to display information to the operator they shall be standardized for meaning within the system and among systems having similar operational requirements.

1.7.4.4 Task Complexity   Software shall minimize operator task complexity. Control inputs shall be simplified to the extent possible, particularly for tasks requiring real-time responses. In addition, control inputs shall permit logical task sequences with a minimum number of control manipulations to achieve task completion.

1.7.4.5 Memorization   The requirements for the operator to memorize system mnemonic codes, special or long sequences, and special instructions must be minimized.

1.7.4.6 System Information   The software shall minimize the requirement for an operator to enter information that is already available to the system. Software should be developed in such a way that once information or data is entered it should not have to be entered again.

1.7.4.7 Labeling   Each individual data group or message shall contain a descriptive title, phrase, word, or similar device to designate the content of the group. Labeling should conform to the following:

- Labels should be unique to prevent operator confusion.
- Labels should reflect the question being posed to the operator when a list of operator options are presented.

1.7.4.8 Natural Language   The language must be natural from the standpoint of the application and the job. Do not use jargon or pejorative terms.

1.7.4.9 Display Functions   The content of a "pop-up" field must be removed when its presence becomes confusing or would be inconsistent with the rest of the screen.

1.7.4.10 Error Messages   Messages must be neither punitive nor humorous.

2 MOUSE (IF REQUIRED)

A mouse shall be used in situations where many selections may be required, or where data may be entered into several fields. *The user* shall *be trained on the proper use of the mouse to avoid problems related to repetitive motion injuries.*

2.1 Buttons

The mouse shall have two buttons for the left and right 'click' capabilities.

2.2 Tactile Feedback

The buttons shall provide the operator with tactile feedback.

2.3   Force

The force for clicking the buttons shall be between 0.9 oz and 2.0 oz.

2.4   Accuracy

An 'Intellimouse' shall be used to optimize accuracy for target acquisition. An optical mouse should be used to reduce tracking errors.

2.5   Tracking Speed

The pointer tracking speed should be adjustable by the user.

2.6   Pointer size/type

The pointer size and type should be adjustable by the user.

2.7   Flexibility

The design of the mouse shall accommodate the left-handed and the right-handed users.

2.8   Hand Shape

The mouse shall have a curved top surface to fit the palm of the hand.

2.9   Dimensions

The maximum length of the mouse shall be 5 inches and the maximum width of the mouse in the narrowest part shall be 2.5 inches.

APPENDIX G
SUPPORTABILITY ENGINEERING CHECKLIST

Author's note—This checklist is provided as an aid for persons responsible for supportability engineering to ensure that applicable areas are being addressed during system acquisition. It should be used as a series of questions, such as "have all system functions been identified using a functional block diagram?" "Have critical system functions been identified?" "Has redundancy been applied to improve mission reliability?"

1. System functions
   a. functional block diagram
   b. critical functions
   c. redundancy

2. Mission
   a. tactical scenario
   b. rate of use
   c. measurement base
   d. systems per mission
   e. mission phases
   f. mission phase code
   g. minimum capability requirement

3. Interoperability
   a. operation
   b. operation support
   c. support
   d. interface control definition

4. Measurement targets
   a. operational effectiveness
   b. operational availability
   c. cost of ownership

5. System life
6. Technology life
7. Situation and use definition
   a. environment of use
   b. number of systems to be procured
   c. number of systems to be supported
   d. locations
      i. operating locations
      ii. support locations
      iii. training locations
      iv. storage locations

    e. deployablility/mobility requirements
        i. operation
        ii. support
        iii. duration
        iv. transport modes
    f. support infrastructure
        i. maintenance and support philosophy
        ii. supply system
            1. wholesale supply support system
            2. retail supply support system
        iii. personnel trade structure
            1. number of personnel available
            2. training methods
            3. training locations
        iv. support and test equipment
            1. standard tool set
            2. standard tool assignment
            3. support equipment list
            4. test equipment list
            5. calibration capabilities
        v. facilities
            1. operation facilities
            2. maintenance facilities
            3. training facilities
            4. storage facilities
            5. special facilities
        vi. PHS&T capabilities
        vii. technical information
            1. operation
            2. maintenance
            3. support

8. Reliability
    a. MTBF
    b. MTBCF
    c. FMECA
    d. criticality analysis
    e. reliability allocation
    f. reliability prediction
        i. preliminary
        ii. final
    g. environmental stress screening
    h. growth test
        i. qualification test

9. Maintainability
    a. MTTR
    b. maintainability allocation
    c. maintainability prediction
    d. maintainability demonstration
        i. plan
        ii. method
        iii. report

10. Testability
    a. fault detection rate
    b. fault isolation rate
    c. fault coverage rate
    d. functional partitioning

11. Reliability centered maintenance
    a. analysis method
    b. FMECA linkage
    c. failure severity
    d. task identification
    e. result disposition
12. Safety
    a. hazard matrix
    b. design assessment
    c. support assessment
    d. hazard log
13. Human factors
    a. person physical limits
    b. access envelope
    c. assessment criteria
    d. LRU limits
14. Availability
    a. inherent availability target
    b. achieved availability target
    c. operational availability target
        i. mission model
        ii. scenario
        iii. limiting factors
        iv. assumptions
15. Cost of ownership
    a. life cycle cost model
    b. through life cost model
    c. whole life cost model
    d. program constants
    e. significant modeling variables
    f. list of modeling assumptions
16. Functional supportability characteristics
    a. standardization
        i. personnel
        ii. parts and materials
        iii. facilities
        iv. tools
        v. support equipment
        vi. test equipment
        vii. technical documentation
        viii. training
        ix. support processes
    b. baseline comparison system
        i. significant comparison issues
        ii. unknowns
        iii. modeling application
    c. accessibility
    d. testing
    e. ease of maintenance
    f. Cost to maintain
    g. Safety to maintain
    h. Interface with existing support systems

17. Physical supportability characteristics
    a. maintenance significant item definition
    b. maintenance task definition
    c. list of maintenance significant items
    d. maintenance task analysis process
    e. resource linkage
    f. task frequency
    g. consolidation of requirements
    h. level of repair analysis
    i. design assessment
    j. validation
       i. plan
       ii. report

18. Obsolescence
    a. design to avoid obsolescence
    b. access to documentation
    c. sources of supply
    d. sources of raw materials

19. System engineering
    a. management plan
    b. implementation plan
    c. design schedule
       i. preliminary design review
       ii. critical design review
       iii. functional configuration audit
       iv. physical configuration audit
    d. testing schedule

20. Design engineering
    a. design schedule
    b. design instructions
    c. supportability assistance
    d. design assessment

21. Purchasing/procurement
    a. procurement schedule
    b. subcontracting
    c. purchasing instructions
    d. supportability assistance

22. Supportability assessment
    a. assessment plan
    b. schedule
    c. participants
    d. report
    e. after action responsibility

23. In-service
    a. mission analysis
       i. system usage rate
       ii. use environment
    b. resource usage rates
       i. materials
       ii. personnel
       iii. facilities
       iv. support equipment
    c. trend analysis

24. Next system planning
    a. mission
    b. environment
    c. technology
    d. obsolescence

25. Lessons learned
    a. report preparation
    b. report distribution

APPENDIX H
CONTRACTING FOR SUPPORTABILITY ENGINEERING

This appendix provides a generic statement of work for performance of supportability engineering activities on an acquisition program. The contents of this appendix must be tailored appropriately for use on an actual project.

1 SUPPORTABILITY ENGINEERING

1.1 The Contractor shall implement a Supportability Engineering program for the Project X program

Continued implementation of supportability engineering analyses commenced by the Contractor prior to Contract Award shall be reported on as updates on information already provided to the Customer.

1.2 Within this section of the Statement of Work (section 1.3), the following definitions apply:

1.2.1 Existing Design.   Design that is the same as that for the original purchaser contract.

1.2.2 New Design.   Design that is completely different from that for the original purchaser contract.

1.2.3 Modified Design.   Design that is a modification of that for the original purchaser contract.

1.2.4   Where the Contractor's proposed design for the Project X Program consists of Existing Design where a Sub-Contractor is performing Supportability Engineering, or has performed Supportability Engineering, on the Existing Design for the Original User, the Contractor shall implement a Management process to assure that the Project X Program requirements will be achieved regardless of changes to the Original user's requirements which may occur after Project X contract award. This Management process shall be to the level of detail necessary to provide the Customer with complete visibility and monitoring of Sub-Contractor Supportability Engineering activities.

1.2.5   Where the Contractor has proposed an Existing Design for the Project X Program, if at any time after Project X Contract Award the Existing Design is shown to be unable or incapable of meeting Project X performance or supportability specifications detailed in the contract, or the Supportability Engineering Program performed by a Sub-Contractor for the Original User is inadequate to meet Project X logistics support requirements detailed in the contract or the Customer Project X Maintenance Policy, the Contractor shall re-categorize the design as either New Design or Modified Design for supportability and shall implement those Supportability Engineering Tasks and Sub-Tasks identified in section 1. herein for that category of design.

1.3 Supportability Engineering Implementation Plan

1.3.1 Sub-Task 1. The Contractor shall prepare and submit a Supportability Engineering Implementation Plan (SEIP) (as an integral portion of the systems engineering plan) as directed by the Customer 30 calendar days after Contract Award.

1.3.1.1 Purpose. The purpose of the SEIP is to provide a management structure and guide for all Supportability Engineering activities of the Project X program. The SEIP shall provide specific information and detail as to all activities to be conducted by the Contractor.

1.3.1.2 Contents. The SEIP is to state fully the requirements for all deliverable data necessary for the Contractor to complete the Project X Contract. The SEIP shall, for each equipment supplied, specifically identify the Supportability Engineering Tasks and Subtasks to be performed by each Project X Sub-Contractor as detailed by the requirements of this Statement of Work and their schedule of performance to meet the Contractor's delivery requirements. The SEIP shall, for each equipment supplied, identify Supportability Engineering Tasks and Subtasks being performed by Project X Sub-Contractors under the requirements of the Original User procurement program which will be used by the Contractor to meet the requirements of this Statement of Work and the schedule of their performance to meet the Contractor's delivery requirements.

1.3.2 Sub-Task 2. Update. Subsequent updates to the SEIP shall be submitted 15 calendar days prior to the start of the Engineering Reviews (Engineering Reviews are detailed in paragraph 1.4).

1.4 Program and Engineering design Reviews

1.4.1 Sub-Task 1. Engineering Design Review Procedures. The Contractor shall establish and document design review procedures which provide for official review and control of released information with Supportability Engineering program participation.

1.4.2 Sub-Task 2. Engineering Design Reviews. Supportability Engineering issues and concerns shall be included in the agendas of all Project X project design reviews conducted for the Project X Project. At a minimum the supportability characteristics and assessment of all new or modified designs shall be addressed for:

1.4.2.1 Project X System to systems, subsystems, and equipment levels;
1.4.2.2 Support Equipment.

1.4.3 Sub-Task 3. Program Reviews. Supportability Engineering issues and concerns shall be included in the agendas of all Project X project program reviews conducted for the Project X Project.

1.5 System Situation and Use Study

1.5.1 A System Situation and Use Study has been prepared by the Customer and shall remain under their control, however, the Contractor shall provide recommended changes to the System Situation and Use Study to maintain an accurate description of the operation, support, and sustainability of the Project X System and Support Equipment. Recommended changes to the System Situation and Use Study shall be included in the agenda of each Engineering Review and shall be delivered to the Customer 15 calendar days prior to the Engineering Review. Recommended changes shall be documented in the form of change pages and shall be submitted under cover letter to the Customer Project X Manager.

1.6   Standardization

1.6.1 Sub-Task 1.   Resource Standardization. The Contractor shall prepare a list of possible standardization candidates for consideration in the design of the Project X system. The possible support resources identified in the Situation and Use Study shall form the basis for this analysis.

1.6.2 Sub-Task 2.   Existing design Analysis. The Contractor shall analyze the support resource requirements for the existing design of the Project X System and Support Equipment and other materials to identify standardization issues which must be addressed during the development of the support package for the Project X Program. Emphasis will be placed on identification of risk due to differences between standardization requirements placed on the original design of the Project X System and Support Equipment and the requirements for standardization with the Customer's existing support structure contained in the Project X Program System Situation and Use Study. The Contractor shall develop and recommend alternatives for resolution or avoidance of standardization problems.

1.6.3 Sub-Task 3.   New or Modified Design.
   1.6.3.1   The Contractor shall identify, develop and implement standardization requirements for any new or modified designs for the Project X System or any Support Equipment developed for the Project X Program.

1.6.4 Reporting.   The Contractor shall identify and provide standardization requirements and issues as an agenda item at each Engineering Review.

1.7   Comparative Analysis

1.7.1 Existing Design
   1.7.1.1   Sub-Tasks. No Sub-Tasks are required.
   1.7.1.2   The Contractor shall not perform a comparative analysis for existing designs of the Project X System and Support Equipment.

1.7.2 New or Modified Design
   1.7.2.1 Sub-Task 1 Baseline Comparison System.   The Contractor shall develop a Baseline Comparison System (BCS) for the Project X System.
   1.7.2.2 Sub-Task 2 Comparison Analysis.   The Contractor shall perform a comparative analysis for new or modified designs of the Project X System and Support Equipment. The existing Project X System and Support Equipment shall be used as the comparative baseline system for the analysis whenever possible. When the existing Project X System is not comparable, the Customer shall approve the comparative system/equipment.

1.7.3 Reporting.   The Contractor shall identify and provide comparative analysis results as an agenda item at each Engineering Review.

1.8   Technological Opportunities

1.8.1 Existing Design
   1.8.1.1 Sub-Tasks.   No Sub-Tasks are required.
   1.8.1.2   The Contractor shall not perform a technological opportunities analysis for existing designs of the Project X System and Support Equipment.

1.8.2 New or Modified Design

1.8.2.1 Sub-Task 1.   Technology Identification. The Contractor shall identify all new and emerging technologies that provide potential improvement in system operational effectiveness, operational availability or cost of ownership.

1.8.2.2 Sub-Task 2.   Technological Opportunities Analysis. The Contractor shall perform a technological opportunities analysis for new or modified designs of the Project X System and Support Equipment. The analysis shall identify technology applications that enhance operation or support of the Project X Program. The analysis shall also identify any risks associated with adoption of a new technology.

1.8.3 Reporting.   The Contractor shall identify and provide technological opportunities as an agenda item at each Engineering Review.

1.9   Supportability Design Criteria

1.9.1 Existing Design.   The supportability design criteria of existing Project X System and Support Equipment shall be in accordance with the off-the-shelf requirements of the Project X Program.

1.9.2 New or Modified Design

1.9.2.1 Sub-Task 1.   Design Objectives. The Contractor shall develop measurable supportability criteria for the Project X system. These criteria shall form the basis for use throughout the operational life of the system to measure the attainment of project goals, thresholds and constraints.

1.9.2.2 Sub-Task 2.   Requirements Implementation. The Contractor shall develop and implement supportability design criteria for each new or modified design of the Project X System and Support Equipment. Each criteria shall be stated as a goal, threshold or constraint. The methods of attainment and the measurement thereof shall be a key measure of success for the systems engineering process.

1.9.3 Reporting.   The Contractor shall present supportability design criteria compliance at each Design and Program review and demonstrate compliance with the criteria through analysis or demonstration. Supportability design criteria shall be an agenda item at each Engineering Review.

1.10   Functional Requirements Identification

1.10.1 Existing Design

1.10.1.1 Sub-Tasks.   Sub-Tasks are required as applicable.

1.10.1.2   The Contractor shall not perform a functional requirements identification for existing designs of the Project X System and Support Equipment for which this analysis was previously performed. If the analysis has not been previously performed, or only partly performed, then the design is to be considered as New or Modified Design for the purposes of maintenance task analysis and the sub-task applicability is to be adequate to enable completion of this task.

1.10.2 New or Modified Design

1.10.2.1 Sub-Tasks

1.10.2.2 Sub-Task 1.   Functional Requirements. The Contractor shall perform a functional requirements identification for each repairable or maintenance significant item on all new or modified designs of the Project X System and Support Equipment, and on existing Project X System and Support Equipment design for which task identification was not previously performed prior to Contract Award. The results of this analysis shall be recorded in an appropriate information system and shall form the basis for any maintenance task analysis performed on the Project X

System and Support Equipment. The analysis must be traceable to task requirements identified as a result of the FMECA, or RCM processes.

1.10.2.3 Sub-task 2.   Design Improvements. The contractor shall identify and implement any results of the functional requirements analysis that improves operational availability or cost of ownership.

1.11   Support System Alternatives

1.11.1 Sub-Tasks.   Sub-Tasks 1. Support System Analysis. The Customer's concept for the support system for the Project X Program is documented in the Project X Situation and Use Study. The Contractor shall recommend changes or modification to the system support concept based on the results of the Supportability Engineering process. The changes shall be presented at Engineering Reviews.

1.11.2 Sub-Task 2.   Support System Analysis. The Contractor shall analyze all components of the system to determine the appropriate support solution that optimizes overall cost of ownership.

1.12   Evaluation and Trade-off Analysis

1.12.1 Sub-Task 1.   Evaluation and Trade-off Process. The Contractor shall establish an evaluation and trade-off process to manage and control supportability and support decisions. The process shall be described in the Contractor's Supportability Engineering Implementation Plan and shall be approved by the Customer.

1.12.2 Sub-Task 2.   Support System Alternatives. The Contractor shall perform a support system alternatives analysis for the Project X Program. The analysis shall be based the Project X System Situation and Use Study, the design and support characteristics of the Project X System, and the existing support requirements and support structure of the original development of the Project X System. The results of this analysis shall be used in the development of maintenance policies and the Maintenance Plan.

1.12.3 Sub-Task 3.   Design, Operation and Support Alternatives. The Contractor shall conduct a design, operation, and support alternatives analysis to develop a recommended support policy and structure for the Project X System. This analysis shall be performed in conjunction with the support system alternatives analysis to determine the most cost effective support policy and structure for the Project X System.

1.12.4 Sub-Task 4.   Readiness Sensitivities. The Contractor shall perform a readiness sensitivities analysis for the Project X System to determine the minimum support policy necessary to provide the necessary support to the Project X System for attainment of readiness requirements. The analysis shall include reliability requirements, maintainability requirements, sparing requirements, manpower and personnel requirements, training and training materials requirements, facility requirements, and other significant resources necessary for the Project X to achieve required operational and mission capabilities.

1.12.5 Sub-Task 5.   Manpower and Personnel Analysis.
1.12.5.1   The Contractor shall perform a manpower and personnel analysis to determine the optimum manning levels and personnel utilization to support the Project X Program. The analysis will focus on minimizing personnel requirements within the approved maintenance policy for the Project X System and Support Equipment and the established manpower budget set by the Customer.
1.12.5.2   The Contractor shall report the estimated manpower requirements at each Engineering Review. The manpower to be reported shall include:

1.12.5.2.1   The total Customer (excluding Project X Project Office staff) manpower and personnel requirements for the Project X Program per quarter for the duration of the Contract.

1.12.5.2.2   Maintenance man-hours (both preventive and corrective) per operating hour at each level of maintenance.

1.12.5.2.3   Estimated maintenance man-hours per maintenance action at each level of maintenance.

1.12.5.2.4   Operational Support man-hours (first line) per war-time mission.

1.12.5.2.5   Operational Support man-hours (first line) per peace-time training mission.

1.12.5.3   The process, methodology and model to be used to derive projected maintenance man hours per operating hour (MMH/OH) shall be included in the Contractor's Supportability Engineering Implementation Plan. These statistics shall also be reported in the monthly Engineering Progress Report. The final manpower and personnel requirements for the Project X Program shall be supported by the approved Project X System and Support Equipment maintenance policy.

1.12.6 Sub-Task 6.   Training Needs Analysis. The Contractor shall optimize the training requirements for the Project X Program. Training requirements shall at minimum address the operation and maintenance of Project X System and Support Equipment.

1.12.6.1   The Contractor may use as the basis for the analysis:

1.12.6.1.1   The Customer provided Training Needs Study (This Study is to be used at the Contractors sole risk).

1.12.6.1.2   The training requirements identified in the analysis of the Project X System and Support Equipment design.

1.12.6.2   The results of this analysis shall be used in the development and preparation of all training courses for Project X operation and maintenance personnel.

1.12.7 Sub-Task 7.   Repair Level Analysis (RLA). The Contractor shall perform a RLA in accordance with a Customer approved specification on the Project X System and maintenance significant Support Equipment to determine the most cost effective method of performing maintenance. The Contractor shall use existing RLA data as a baseline for performance of this analysis. However, when the RLA decision criteria differ from those agreed with the Customer, then the RLA shall be performed with the Customer's criteria. The Contractor shall recommend in the Supportability Engineering Implementation Plan the most appropriate RLA software model for performance of this task for Customer approval. The Contractor shall provide a copy of the RLA software with all documentation required to operate the software, to the Customer. The results of this analysis shall be presented at Engineering Reviews and shall be an integral part of recommended changes to the Project X System Situation and Use Study.

1.12.8 Sub-Task 8.   Diagnostics and Testing. The Contractor shall perform an analysis to determine the optimum approach for maintenance of the Project X System and Support Equipment in the use of BIT and support and test equipment for diagnostics and testing of the System. The results of this analysis shall be incorporated into the repair level analysis process to ensure that implementation of diagnostics and testing fully supports achievement of optimum mean man-hour per flying hour and mean man-hour per maintenance action.

1.12.9 Sub-Task 9.   Comparative Evaluations. The Contractor shall perform comparative evaluations of existing Project X use and maintenance to optimize the Project X System maintenance policy, support requirements, and introduction into service. The results of this analysis shall be used to support finalization of the Project X System maintenance and support polices.

1.12.10 Sub-Task 10.   Energy Trade-Offs. The Contractor shall evaluate and trade-off Project X system/equipment alternatives and energy requirements.

1.12.11 Sub-Task 11.   Survivability Analysis. The Contractor shall perform a survivability analysis of the Project X to develop battle damage repair requirements. The results of this analysis shall be used to identify and develop battle damage repair procedures.

1.12.12 Sub-Task 12.   Transportability Analysis. The Contractor shall perform a transportability analysis for the Project X System, Support Equipment, and support resources. The results of this analysis shall be used to verify packaging, handling, storage and transportability (PHS&T) requirements.

1.12.13 Sub-Task 13.   Support Facility Trade-Offs. The Contractor shall evaluate and trade-off Project X system/equipment alternatives and support facilities requirements. This analysis shall focus on minimizing requirements for support resources.

1.12.14 Life Cycle Cost.   The Contractor shall perform a logistics life cycle cost (LCC) analysis for the Project X Program. The LCC shall support overall evaluation and trade-off analyses performed as part of Supportability Engineering. The Contractor shall propose a recommended LCC software package for performance of this requirement. The recommended LCC software package shall be identified in the Contractor's SEIP. The Contractor shall provide to the Project X Program, at no cost, a complete LCC software package including training and documentation. The Contractor shall provide at each Engineering Review the estimated LCC for the Project X Program based on the results of the Supportability Engineering process.

1.12.15 Reporting.   The Contractor shall include each Trade-off analysis as an agenda item for each Engineering Review.

1.13   Maintenance Task Analysis

1.13.1   The Contractor shall perform a maintenance task analysis for each maintenance task requirement identified on the Project X System and Support Equipment.

1.13.2 Sub-Tasks.   All Sub-Tasks are required.

1.13.3 Sub-Task.   1 shall specifically include:
  1.13.3.1   Sequential steps to perform the maintenance task,
  1.13.3.2   Personnel skill and elapsed time requirements,
  1.13.3.3   Support Equipment required,
  1.13.3.4   Facility requirements,
  1.13.3.5   Training task requirements,
  1.13.3.6   Spares, parts, and consumable requirements,
  1.13.3.7   Human Factors Engineering requirements,
  1.13.3.8   Safety limitations or requirements,
  1.13.3.9   Task frequency,
  1.13.3.10   Transportability and PHS&T requirements,
  1.13.3.11   Maintenance level,
  1.13.3.12   Security requirements.

1.13.4   The results of the maintenance task analysis shall be recorded in a Supportability Engineering Database.

1.13.5   New or Critical Resources identified through task analysis shall be reported at each Engineering Review.

1.14   Early Fielding Analysis

1.14.1   The Contractor shall assist the Customer in conducting an early fielding analysis for the Project X Program. The analysis shall focus on potential support problems encountered when the Project X system is fielded. Additionally, the analysis shall address anticipated problems the Customer may encounter with the overall support infrastructure when the Project X is fielded. The results of this analysis shall be provided at each Engineering Review.

1.15 Post-Production Support Analysis

1.15.1 The Contractor shall assist the Customer in conducting a post production support analysis for the Project X Program. The analysis shall focus on projected obsolescence problems for the Project X system and the methods used to resolve or avoid anticipated post production support problems. The results of this analysis shall be provided at each Engineering Review.

1.15.2 The Contractor shall initiate a diminishing manufacturing sources and materials shortage (DMSMS) program to limit the effect of obsolescence. The DMSMS program will be maintained for the duration of this contract.

1.16 Supportability Test, Evaluation and Verification

1.16.1 The Contractor shall conduct a supportability Test, Evaluation and Verification (TEV) program for the Project X System and Support Equipment. Verification of equipment is to be in accordance with the TEA Plan and the TEV Plan.

1.16.2 Supportability Qualification and Acceptance. The Contractor shall demonstrate the supportability of the final design of the Project X System as part of the comprehensive Qualification and Acceptance process as defined in the Q&A procedures for the Project X Program.

1.16.3 Supportability TEV Plan. The Contractor shall prepare and submit a Supportability TEV Plan as directed by the Customer 30 calendar days after Contract Award. The plan at a minimum shall contain identification of the TEV process, resources required to support TEV, and a schedule of TEV activities. The TEV plan shall include the complete procedure, resources, and methodology to be used in qualification and acceptance testing of the operational and supportability characteristics of the Project X System and Support Equipment. Each operational and support task for the Project X System shall be TEV through analysis or demonstration. Maintenance tasks shall be TEV through demonstration; at which time, all of the resources required to perform the maintenance tasks shall be verified. The TEV Plan shall also include requirements for in service monitoring of Project X System and Support Equipment. The TEV Plan shall also include validation of the information and associated data used to develop the physical support solution for the system.

1.16.4 Supportability TEV Report. The Contractor shall prepare and submit a supportability TEV report as directed by the Customer 30 calendar days after the completion of the TEV. The report shall document the results of all supportability TEV of the Project X System.

1.16.5 Status Reporting. The Contractor shall include supportability TEV as an agenda item for each Engineering Review.

1.16.6 Supportability demonstration/Maintainability Demonstration. The final TEV of the Project X System and Support Equipment design and support requirements shall consist of a Supportability demonstration/Maintainability Demonstration (SD/MD) conducted by the Customer as an integral part of the system performance testing. The SD/MD shall be conducted at Contractor premises. The Contractor shall provide production standard (i.e. representative of that to be delivered to the Customer) Project X System, Support Equipment to the agreed program schedule. The results of the SD/MD will be provided to the Contractor by the Customer 30 calendar days after completion of the SD/MD. The Contractor shall institute corrective actions required to eliminate or correct any deficiencies identified by the SD/MD. Should corrective action require Project X System and Support Equipment redesign and/or revision to support data, such redesign and/or revision shall be at no cost to the Customer. The corrections shall be implemented prior to the Scheduled delivery to the Customer of the equipment unless otherwise agreed by the Customer.

1.16.6.1 The Contractor shall prepare and submit a SD/MD Plan for the Customer's approval thirty calendar days after Contract Award. The Plan shall contain at a minimum:

1.16.6.1.1 Schedule for conduct of the SD/MD

1.16.6.1.2 Methodology for performance (implementation of the applicable standards)

1.16.6.1 Customer participation requirements

1.16.6.2 The SD/MD shall enable the Customer to:

1.16.6.2.1 Verify ease of use.

1.16.6.2.2 Verify the assignment of maintenance tasks.

1.16.6.2.3 Verify that all training equipment is justified, accurate and effective.

1.16.6.2.4 Verify that all training courses are justified, accurate and effective.

1.16.6.2.5 Measure contractually agreed times for engine and major assembly replacements, corrective maintenance actions, scheduled maintenance actions, and servicing.

1.16.6.2.6 Verify the content of corrective maintenance actions and scheduled maintenance actions.

1.16.6.2.7 Verify ease of maintenance.

1.16.6.2.8 Verify daily servicing and technical servicing.

1.16.6.2.9 Verify start-up procedures and test schedules.

1.16.6.2.10 Verify the requirement for, and suitability of, system calibration procedures.

1.16.6.2.11 Verify built-in test (BIT) performance.

1.16.6.2.12 Verify all Test Program Sets.

1.16.6.2.13 Verify all rigging procedures.

1.16.6.2.14 Verify all support and test equipment requirements as fully identified, justified and suitable.

1.16.6.2.15 Verify the content delivery medium of publications to ensure that they are fit for their intended purpose and use, that they are technically accurate and that they are safe in application and the interactive electronic technical manual (IETM) delivery system is reliable and easy to use.

1.16.6.2.16 Verify the performance of tasks in clothing/dress states as defined in the Situation and Use Study.

1.16.6.2.17 Verify the performance of tasks in defined environmental conditions, light levels, snow, slopes etc., as defined in the Situation and Use Study.

1.16.6.2.18 Confirm that the Customer's manpower and personnel skill requirements are correct, and that any new skills that may be required have been identified and are justified.

1.16.6.2.19 Verify the requirement and justification for any mobile/fixed maintenance facilities.

1.16.6.2.20 Verify the acceptability of the proposed systems for the transportation and recovery of the system by road and as an internal or under slung load in transport aircraft and by ship.

1.16.6.2.21 Materiel Handling Equipment. Verify the suitability of and justification for all items of proposed Materiel Handling Equipment.

1.16.6.2.22 Special to Type Vehicles (STV). Verify the suitability of and justification for all STV.

1.16.6.2.23 Special Test Equipment. Verify the suitability of the system capability of any special test equipment.

1.16.6.2.24 Special to Contents PHS&T Requirements. Verify all Special to Type PHS&T requirements, procedures and equipment (other than Material Handling Equipment).

1.16.6.3 The Contractor shall demonstrate, as requested by the Customer, any task which the Customer considers to be hazardous to the system or its own personnel.

1.16.6.4 The Contractor shall prepare and submit a report on the proposed implementation of the results of the SD/MD for the Customer's approval.

2 SUPPORTABILITY ENGINEERING DATABASE

2.1 The Contractor shall prepare and maintain a Supportability Engineering Database for the Project X Program

The Contractor shall utilize Supportability Engineering Database validated to comply with the full requirements of GEIA-STD 0007, Logistics Product Data, MIL-STD 1388-2B, Logistics Support Analysis Record, Def Stan 00-60, Integrated Logistics Support, or a comparable internationally recognized standard.

2.2 Existing Data

The Contractor shall prepare Supportability Engineering data for existing designs of the Project X System and Support Equipment. The database shall be prepared using existing source data. The Contractor shall ensure, at no cost to the Customer, that the existing source data reflects the current design and support requirements of the Project X System.

2.3 New or Modified Design

The Contractor shall prepare Supportability Engineering data for new or modified designs or support requirements for the Project X System and Support Equipment. The Supportability Engineering Database data shall record the results of the maintenance task analysis and shall reflect the decision made through trade-off analysis of system support options.

2.4. Supportability Engineering Database Use

The Contractor shall use the Project X Program Supportability Engineering Database as the basis for all recommended support for the Project X System, and for source data in preparation of applicable final versions of deliverable support documentation.

2.4.1 The Customer shall be advised of the progressive population of the Supportability Engineering Database to support the following sequence of Customer activities:

2.4.1.1 Maintenance Significant Items. The Supportability Engineering Candidate Item List (CIL) shall reside in the Supportability Engineering Database. The Customer shall have the capability of extracting the Supportability Engineering CIL from the Supportability Engineering Database.

2.4.1.2 Reliability, Maintainability, Testability, Availability, Reliability Centered Maintenance analysis performed by the Contractor shall be documented in the Supportability Engineering Database. The results of these analyses shall be used by the Contractor in the identification of Supportability Engineering Candidates and Maintenance Tasks.

2.4.1 Maintenance Task Identification. The next step in development of the Supportability Engineering Database is identification of all maintenance tasks required for each Supportability Engineering Candidate. The Contractor shall perform task identification. The Customer shall have the capability of extracting the maintenance tasks identified for each Supportability Engineering Candidate from the Supportability Engineering Database.

2.4.1.4 Maintenance Task Analysis (MTA). The first step of the MTA shall be to perform a limited analysis to determine the resources required to perform the task. Resources identified as part of the limited MTA include parts, spares, tools, support equipment, test equipment, personnel (numbers, trades, skills), task time, and facilities. This first step shall not necessarily include a detailed writing of the maintenance task steps. The important contents of this step of the MTA shall be to assure that the maintenance task is done properly and the correct resources have been identified to support of task performance. The results of the limited MTA shall be extracted from the Supportability Engineering Database.

2.4.1.5 Level of Repair Analysis. At the completion of the limited MTA, the next step shall be to determine which maintenance tasks will actually be done and where they will be done. This shall be accomplished through Level of Repair Analysis (LORA). The Contractor shall perform LORA. There are two steps to LORA, non-economic analysis and then economic analysis. Non-economic analysis consists of analyzing each maintenance task to determine if there is some non-economic reason which would dictate not performing a task or would dictate a set repair policy in accordance with satisfying any Customer imposed Non-Economic Pre-Empting Factor. Typical non-economic analysis is driven by technology, safety, limited capabilities, or policy. If there is no non-economic reason for a LORA decision, then the Contractor shall perform an economic LORA to determine the most cost effective maintenance policy for an item. The LORA results are recorded in the Supportability Engineering Database. The Customer shall have the ability to extract the LORA results from the Supportability Engineering Database.

2.4.1.6 Detailed Maintenance Task Analysis. At the completion of the LORA and approval by the Customer, the Contractor shall then perform a detailed maintenance task analysis which shall result in a complete written maintenance task. The MTA at this stage shall also revisit resource requirements to verify that all resources required to perform the task have been completely identified and documented in the Supportability Engineering Database. The results of the detailed MTA when recorded in the Supportability Engineering Database shall be in the form of "authored text" which is assembled in the Supportability Engineering Data. Additionally, all spares or parts required to perform a maintenance task shall be linked to the Supportability Engineering Analysis; tools and support equipment required to perform a maintenance task shall be linked to the Supportability Engineering Analysis; test equipment required to perform a maintenance task shall be linked to the Supportability Engineering Analysis; personnel required to perform a maintenance task shall be linked to the Supportability Engineering Analysis; and facilities required to support performance of a maintenance task shall be linked to the Supportability Engineering Analysis.

2.4.1.7 Completion of the Supportability Engineering Database.

2.4.1.7.1 Results of Reliability, Maintainability, Testability, Availability, Reliability Centered Maintenance analyses performed by the Contractor shall be documented in the Supportability Engineering Database including support statistics, such as MTBF, MTTR, etc.

2.4.1.7.2 Personnel requirements to support maintenance of the Project X shall have been determined by individual maintenance task analyses, however, total personnel requirements shall be determined by consolidating the results of the individual tasks analyses into a total system level requirement. The Customer will be able to make this consolidation from the Supportability Engineering Database to determine the total manpower requirements for the Project X program. The Customer will have established a program manpower budget, and the Customer shall be able to compare the results of the MTA with the budget to determine any shortfalls or requirements for staffing realignment. The Customer shall also have the ability to identify individual maintenance task that require long or excessive manpower which may be significantly degrading the overall use of personnel in support of Project X maintenance.

2.4.1.7.3 Training requirements for maintenance personnel shall be based on the results of the MTA. As the individual MTAs are completed, the Contractor's training experts shall review each maintenance task to determine if any special or new training is required for Customer personnel to perform the task. If so, the Supportability Engineering Data for the task shall be annotated to signify such a requirement. Where additional training is recommended, the Contractor's training development team shall annotate the specific training requirements for each maintenance task. The Customer shall be able to extract consolidated training requirements from the Supportability Engineering Database.

2.4.1.7.4 Support Equipment, Tools, and Test Equipment requirements identification shall be established by the MTA. This shall be used by the Contractor in development of lists of support equipment, tools, and test equipment which will be required to be procured and positioned to support maintenance. Additional information about each item may be required for procurement or detailed authorization and shall be entered by the Contractor in Supportability Engineering Data as appropriate. The Customer shall be able to review the total requirements for these items.

2.4.1.7.5   Spares Provisioning should be logical progression from the MTA. After items have been identified by the MTA, Contractor provisioning experts shall continue development of all provisioning data by completion of the appropriate information on the Supportability Engineering Data. Contractor shall produce all Initial Provisioning from the Supportability Engineering Database. The Customer shall extract spares and provisioning data from the Supportability Engineering Database and verify requirements for the item.

2.4.1.7.6   Facilities requirements for the Project X program shall be consolidated on the Supportability Engineering Data shall be linked to each MTA. The Supportability Engineering database shall allow the Customer to review the total facility impact of the Project X program.

2.4.1.7.7   PHS & T requirements resulting from support or operation of the Project X shall be recorded in Supportability Engineering Data. This will include, but not be limited to, major items requiring transport, spares and repair parts which are to be packaged for storage and shipment and support equipment. The Contractor shall be responsible for developing a cost effective scheme for Project X PHS & T through life.

2.5 Supportability Engineering Summary Reports

The Contractor shall provide Supportability Engineering summary reports on an as-needed basis for Engineering Reviews.

2.6 Delivery

The Contractor shall prepare and submit an initial copy of the Supportability Engineering Database 15 calendar days prior to the first Supportability Engineering Review after Supportability Engineering Database preparation commences. Thereafter, changes, modifications, additions, and deletions shall be delivered 15 calendar days prior to each Engineering Review. A final copy of the complete, validated Supportability Engineering Database shall be delivered 60 calendar days prior to end of contract. The media for each delivery shall be agreed prior to Contract Award; however, the Customer prefers all deliveries through electronic media of CD or encrypted e-mail.

2.7 Initial Provisioning

The Contractor shall prepare initial provisioning requirements from the Supportability Engineering Database. The Contractor shall not maintain separate Supportability Engineering Database and provisioning databases.

2.8 Update

The Contractor shall maintain and update the Supportability Engineering Database throughout the Contract.

INDEX